Audio Post Production for Television and Film

Audio Post Production for Television and Film

An introduction to technology and techniques

Third edition

Hilary Wyatt and Tim Amyes

AMSTERDAM · BOSTON · HEIDELBERG · LONDON · NEW YORK · OXFORD
PARIS · SAN DIEGO · SAN FRANCISCO · SINGAPORE · SYDNEY · TOKYO
Focal Press is an imprint of Elsevier

Focal Press
An imprint of Elsevier
Linacre House, Jordan Hill, Oxford OX2 8DP
30 Corporate Drive, Burlington, MA 01803

First published as the *Technique of Audio Post-Production in Video and Film* 1990
Paperback edition 1993
Reprinted 1994, 1995 (twice), 1997, 1998
Second edition 1998
Reprinted 2000
Third edition 2005
Reprinted 2005

British Library Cataloguing in Publication Data
A catalogue record for this book is available from the British Library

Library of Congress Cataloguing in Publication Data
A catalogue record for this book is available from the Library of Congress

ISBN 0 240 51947 7

For information on all Focal Press publications
visit our website at www.focalpress.com

Working together to grow
libraries in developing countries

www.elsevier.com | www.bookaid.org | www.sabre.org

ELSEVIER   BOOK AID International   Sabre Foundation

Typeset by Newgen Imaging Systems (P) Ltd., Chennai, India
Printed and bound in Great Britain

Contents

Acknowledgements

Thank you to Dennis Weinreich, Richard Conway and all my good friends at Videosonics for their enthusiasm and support during the writing of this book, and their willingness to share their knowledge and experience with me. I'd especially like to thank Jeremy Price, Simon Gershon, Dave Turner, Michele Woods, Andrew Tyndale, Andrew Stirk, Barnaby Smyth, Howard Bargroff, Dan Johnson and Smudger.

I'd also like to thank the many friends and colleagues within the industry who have been so generous with their time and advice – Simon Bishop, Richard Manton, Tim Alban, Alex Mackie, Kate Higham, Heidi Freeman, Sam Southwick, Thomas Drescher, Roger Knight, Jon Matthews, Ed Bulman, Anthony Faust and Jim Guthrie.

Thanks to Peter Hoskins at Videosonics, Clare MacLeod at Avid, Liz Cox at Digidesign, and Mike Reddick at AMS Neve for their help with many of the illustrations.

Hilary Wyatt

In writing this book many friends and colleagues in the industry have contributed. Among these are Andy Boyle, Ian Finlayson, Cy Jack, Gillian Munro, Mike Varley, Alistair Biggar, Tim Mitchell, Len Southam and many others.

The extract from the sound notes by Alfred Hitchcock for his film *The Man Who Knew Too Much* are by kind permission of the Alfred Hitchcock Trust.

Tim Amyes

About the authors

Hilary Wyatt is a freelance Dialogue Supervisor and Sound Effects Editor. She began her career in 1987, creating sound effects and editing music for a number of long-running British 'cult' animation series.

Since then, Hilary has worked as a Sound Effects Editor on a wide range of productions, including commercials, documentary, drama and feature films. In 1999 she supervised the dialogues on the British gangster film *Sexy Beast*, and has since worked as Dialogue Supervisor on a number of British and American features. Recent credits include *Jojo In the Stars* (animation), *Absolute Power* and *Dr Zhivago* (TV), *Bright Young Things, Dear Frankie, Something Borrowed* and *White Noise* (features).

Tim Amyes has many years experience in post production, covering the whole production chain. As well as being a former sound supervisor at Scottish Television, Tim has worked at two other companies in the British ITV network, both as a sound recordist and dubbing mixer. He has been involved in industry training from the start, serving as one of the original members of Skillset, which was set up to provide training standards (NVQs) for the UK television and film industries. He has also served on both national industry and union training committees, and advised on the recent Scottish Screen/BFI publication *An Introduction to Film Language*.

Currently, Tim lectures in audio, produces specialist corporate videos, and writes, having sold documentary scripts to both the BBC and ITV. A keen film enthusiast, he is a past member of the Scottish Film Archive's advisory committee.

Introduction to the third edition

The focus of this book is audio post production, one of the last stages in the creative process. By the time it takes place, many crucial decisions will have been made regarding the sound – sometimes in consultation with the sound post production team, sometimes not! It is important for those working in audio post production to have a working knowledge of what happens on location, or in the studio, and during the picture edit, as all these stages will have a bearing on their own work, both technically and creatively. The third edition has therefore been completely rewritten and restructured to provide a step-by-step guide to the professional techniques that are used to shape a soundtrack through the production process.

This edition is split into two parts. Part 1 deals with the technical nuts and bolts of audio post production – how audio is recorded, how sound and picture are synchronized together, how audio is transferred between systems, and how film and video technology works. You may find it useful to refer back to these chapters when reading the second part of the book, which follows the path of production sound from its original recording right through to the final mix and transmission. Part 2 is structured to follow a typical post production workflow. It examines the equipment used at each stage, how it is used, and it includes many of the practical techniques and shortcuts that are used by experienced editors and mixers.

This book uses the generic terms 'non-linear picture editor' (abbreviated to NLE) and 'digital audio workstation' (abbreviated to DAW) to describe systems in general use at the current time. On some occasions we have been more specific, and have mentioned actual manufacturers where we felt it important. However, we have only named equipment we feel is in standard use, and which is likely to be around for many years to come. The reader should, however, bear in mind that some aspects of audio technology are changing at a very rapid rate.

Hilary Wyatt

Introduction to the third edition

Part 1

Audio Basics

1 | The evolution of audio post production

Hilary Wyatt

An overview

The term *audio post production* refers to that part of the production process which deals with the *tracklaying, mixing* and *mastering* of a soundtrack. Whilst the complexity of the finished soundtrack will vary, depending on the type of production, the aims of the audio post production process are:

- To enhance the storyline or narrative flow by establishing mood, time, location or period through the use of dialogue, music and sound effects.
- To add pace, excitement and impact using the full dynamic range available within the viewing medium.
- To complete the illusion of reality and perspective through the use of sound effects and the recreation of natural acoustics in the mix, using equalization and artificial reverbs.
- To complete the illusion of unreality and fantasy through the use of sound design and effects processing.
- To complete the illusion of continuity through scenes which have been shot discontinuously.
- To create an illusion of spatial depth and width by placing sound elements across the stereo/surround sound field.
- To fix any problems with the location sound by editing, or replacing dialogue in post production, and by using processors in the mix to maximize clarity and reduce unwanted noise.
- To deliver the final soundtrack made to the appropriate broadcast/film specifications and mastered onto the correct format.

The Man Who Knew Too Much – Reel VII
Alfred Hitchcock sound notes:

The scenes in the hotel room continue with the same sound as indicated above.

In the Camden Town street we should retain something of the suburban characters of the barking dog and the distant hammering iron.

Now to the question of the footsteps. These are very, very important. The taxi that drives away after Jimmy gets out should be taken down as quickly as possible because we want no sounds other than very distant traffic noises because the predominant sound is the footsteps of Jimmy Stewart. They seem to have a strange echo to him because they almost sound like a second pair of footsteps, until he stops to test it and the echoing footsteps also stop. When he resumes, they resume. And to test it further he stops again, but this time the echoing footsteps continue. Then he slows down and the echoing footsteps slow down. Now as he proceeds the echo gets louder but his own footsteps remain the same volume. And when he looks around the second time we see the reason for the echoing foot-steps. They belong to the other man. Now the two sets of echoing footsteps are heard. The quality of echoing footsteps diminishes and they become more normal than Jimmy's and remain normal as the other man passes Jimmy and crosses the street to enter the premises of Ambrose Chappell.

Make sure the bell that Jimmy presses at the Taxidermist's door is rather a weak one and has quite an old-fashioned quality. Don't have it a very up-to-date sharp ring because it would be out of char-acter with the place.

Once Jimmy is in the room there should be just the faint sound of men at work, a cough or two and perhaps a bit of filing noise, an odd tap of a light hammer, etc.

Mr Tomasini has special notes concerning the re-dubbing of Jimmy's lines after Ambrose Chappell Jnr. has said they have no secrets from their employees.

Note that the correct amount of dialling should be heard when Ambrose Sr. dials. When Jimmy opens the door to exit let us hear some outside traffic noise and banging iron noise again and bark-ing dog noise just for a brief moment.

Back in the Savoy suite the same sounds apply but it would be wise to avoid Big Ben again. Otherwise we would be committing ourselves to certain times which we should avoid.

Outside Ambrose Chappell again some suburban noise, distant children's cries at play, and the odd traffic that goes by at the end of the street.

Reel VIII ...

Figure 1.1 Alfred Hitchcock's sound spotting notes for *The Man Who Knew Too Much* (courtesy of The Alfred Hitchcock Trust).

A little history: the development of technology and techniques

Despite the fact that audio post production is now almost entirely digital, some of the techniques, and many of the terms we still use, are derived from the earliest days of film and television production.

The first sound film

The first sound film was made in America in 1927. *The Jazz Singer* was projected using gramophone records that were played *in synchronization* with the picture: this was referred to as the *release print*. Film sound played on the current enthusiasm for radio, and it revived general public interest in the cinema. Around the same time, Movietone News began recording sound and filming pictures of actual news stories as they took place, coining the term *actuality sound and picture*. The sound was recorded photographically down the edge of the original camera film, and the resulting *optical soundtrack* was projected as part of the picture print.

At first, each news item was introduced with silent titles, but it was soon realized that the addition of a commentary could enliven each *reel* or *roll* of film. A technique was developed whereby a spoken *voice-over* could be mixed with the original actuality sound. This mix was copied or recorded to a new soundtrack: this technique was called 'doubling', which later became known as *dubbing*. Any extra sounds required were recorded to a separate film track, which was held *in sync* with the original track using the film sprockets.

Early editing systems

Systems were developed that could run several audio tracks in sync with the picture using sprocket wheels locked onto a drive shaft (see Figure 1.2).

The synchronizer and Moviola editing machines were developed in the 1930s, followed by the Steenbeck. The fact that shots could be inserted at any point in a film assembly, and the overall sync adjusted to accommodate the new material, led to the term *non-linear editing*.

Dubbing/re-recording

Early *mixing consoles* could only handle a limited number of tracks at any one time – each channel strip controlled a single input. Consoles could not play in reverse, nor *drop in* to an existing mix, so complete reels had to be mixed *on-the-fly* without stopping. This meant that tracks had to be *premixed*, grouping together *dialogue, music* and *fx* tracks, and mixes took place in specially built *dubbing theatres*. Each of the separate soundtracks was played from a *dubber*, and it was not unusual for 10 machines to be run in sync with each other. A *dubbing chart* was produced to show the layout of each track.

Early dubbing suffered from a number of problems. Background noise increased considerably as each track was mixed down to another and then copied onto the final print – resulting in poor *dynamic range*.

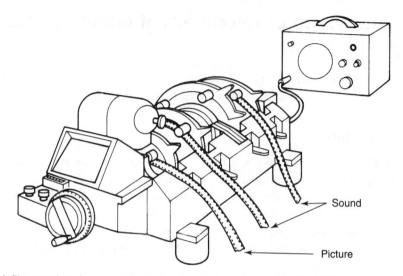

Figure 1.2 A film synchronizer used for laying tracks (courtesy of A. Nesbitt).

The *frequency range* was also degraded by the dubbing process, which made each generation a poor copy of the first. To improve speech intelligibility, techniques were developed which involved modifying the frequency response of a signal. However, the lack of a uniform standard meant that many mixes were almost unintelligible in some poorly equipped theatres. To tackle the problem, The Academy of Motion Picture Arts and Sciences developed the 'Academy Curve' – an equalization standard which ensured that all engineers worked to the same parameters, and which standardized monitoring in dubbing theatres and cinemas. This standard was maintained until the 1970s, when film sound recording was re-appraised by the Dolby Corporation.

By the end of the 1930s, the film industry had refined all the fundamental techniques needed to produce a polished soundtrack – techniques that are still used to some extent today.

Post sync recording

It became standard practice to replace dialogue shot on set with better quality dialogue recorded in a studio. American producers needed a larger market than just the English speaking world, and so *post synchronization* techniques were developed to allow the re-voicing of a finished film in a foreign language. A picture and sound film *loop* was made up for each line. This loop could be played repeatedly, and the actor performed the line until an acceptable match was achieved. A chinagraph line marked on the film gave the actor an accurate *cue* for the start of each loop. This system of *automatic dialogue replacement* or *looping* is still in use today, although electronic beeps and streamers have replaced the chinagraph lines. Footsteps and moves were also recorded to picture, using a post sync technique invented by Jack Foley, a Hollywood sound engineer. This technique is still used on many productions and is known as *foley recording*.

Stereo

The first stereo films were recorded using true stereo to record dialogue on set. Unfortunately, once the footage was cut together, the stereo image of the dialogue would 'move' on every picture cut – perhaps even in mid-sentence, when the dialogue bridged a cut. This was distracting and led to the practice of recording dialogue in mono, adding stereo elements later in post production. This is still standard practice in both TV and film production.

In 1940, Walt Disney's *Fantasia* was made using a six-channel format which spread the soundtrack to the left, right and centre of the screen, as well as to the house left, right and centre channels. This echoed the use of *multichannel* formats in use today, such as Dolby Surround, Dolby Digital and DTS.

The desire to mix in stereo meant an increase in the number of tracks a mixer had to control. Mixing was not automated until the late 1980s, and the dubbing mixer had to rely on manual dexterity and timing to recreate the mix 'live' on each pass. Early desks used rotary knobs, rather than faders, to control each channel. The introduction of *linear faders* in the 1960s meant that mixers could more easily span a number of tracks with both hands. Large feature-film mixes often took place with two or three mixers sat behind the console, controlling groups of faders simultaneously.

Magnetic recording

Magnetic recording was developed in the 1940s and replaced optical recording with its much improved sound quality. Up to this point, sound editors had cut sound by locating modulations on the optical track, and found it difficult to adjust to the lack of a visual waveform. Recent developments in digital audio workstations have reversed the situation once more, as many use *waveform editing* to locate an exact cutting point. Magnetic film was used for film sound editing and mixing right up to the invention of the digital audio workstation in the 1980s.

The arrival of television in the 1950s meant that cinema began to lose its audience. Film producers responded by making films in widescreen, accompanied by multichannel magnetic soundtracks. However, the mag tracks tended to shed oxide, clogging the projector heads and muffling the sound. Cinema owners resorted to playing the mono optical track off the print (which was provided as a safety measure), leaving the surround speakers lying unused.

Television

Television began as a live medium, quite unlike film with its painstaking post production methods. Early television sound was of limited quality, and a programme, once transmitted, was lost forever. In 1956, the situation began to change with the introduction of the world's first commercial videotape recorder (VTR). The Ampex VTR was originally designed to allow live television programmes to be recorded on America's East Coast. The recording would then be time delayed for transmission to the West Coast some hours later. This development had two significant consequences. Firstly, recordings could be edited using the best takes, which meant that viewers no longer saw the mistakes made in live transmissions. Secondly, music and sound effects could be mixed into the original taped sound. Production values improved, and post production became an accepted part of the process.

In 1953 NTSC, the first ever colour broadcast system, was launched in the US based on a 525-line/60 Hz standard. The PAL system, based on a 625-line/50 Hz standard, was introduced across Europe (except for France) in the early 1960s. At the same time, SECAM was introduced in France. This is also a 625-line/50 Hz standard, but transmits the colour information differently to PAL.

Video editing

This was originally done with a blade, and involved physically cutting the tape, but this was quickly superseded by *linear editing*. Here, each shot in the edit was copied from a source VTR to the record VTR in a linear progression. Tape editing is still used in some limited applications, but the method is basically unchanged. However, just as early film soundtracks suffered from generational loss, so did videotape because of the copying process. This problem was only resolved with the comparatively recent introduction of digital tape formats.

Timecode

Sound editing and *sweetening* for video productions was originally done using *multitrack recorders*. These machines could record a number of tracks and maintain sync by *chasing* to picture. This was achieved by the use of *timecode* – a technology first invented to track missiles. Developed for use with videotape in 1967, it identified each individual frame of video with accurate clock time, and was soon adopted as the industry standard by the Society of Motion Picture and Television Engineers (SMPTE). The eight-digit code, which is at the heart of modern editing systems, is still known as *SMPTE timecode*.

Videotape formats

Early video equipment was extremely unwieldy, and the tape itself was 2 inches wide. In the 1970s, 1-inch tape became the standard for studio recording and mastering. Because neither format was at all portable, location footage for television (e.g. news and inserts for studio dramas, soaps and sitcoms) was shot using 16-mm film. This remained the case until the U-matic ¾-inch format was released in 1970. Designed to be (relatively) portable, it marked the beginnings of *electronic news gathering* or *ENG*. Both picture and sound could now be recorded simultaneously to tape. However, it was the release of Sony Betacam, and its successor Betacam SP in 1986, which spelled the end of the use of film in television production, although 16- and 35-mm film continued to be used for high-end drama and documentary. Betacam SP also replaced the use of 1-inch videotape and became the industry standard format until the 1990s.

Dolby Stereo

In 1975, Dolby Stereo was introduced to replace the then standard mono format. In this system, a four-track *LCRS* mix (Left, Centre, Right, Surround) is encoded through a *matrix* as two tracks carrying left and right channels, from which a centre track and surround track are derived. These two channels (known as the *Lt Rt* – Left total Right total) are printed optically on the film edge and decoded through a matrix on playback. Where the playback system does not support surround monitoring, the sound-track will play out correctly as a mono or stereo mix. This type of soundtrack is known as an *SVA* or

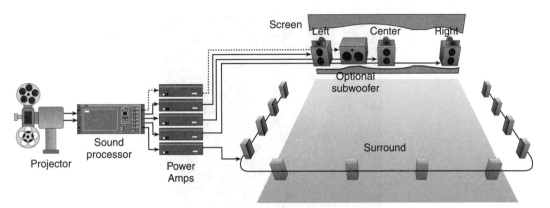

Figure 1.3 Dolby Stereo (courtesy of Dolby Laboratories Inc.).

stereo variable area. The original Dolby Stereo format used existing Dolby A noise reduction technology to reduce the problems associated with optical soundtracks. In 1986 this was replaced by Dolby SR (Spectral Recording), which offered better noise reduction and an improvement in dynamic range. Today, analogue soundtracks for theatrical release are still made using Dolby SR.

Digital audio

In the 1960s and 1970s, a number of digital tape recorders and synthesizers were developed, which could make multi-generational copies without adding noise or distortion. It was not until 1980 that Philips/Sony completed the development of the Compact Disc, which played digitally encoded audio using 16-bit/44.1 kHz pulse code modulation. In 1986 Philips/Sony released the Digital Audio Tape or *DAT*, which would quickly replace ¼-inch analogue reel-to-reel tape as the standard location recording format, and which is only now being replaced by location recorders that store audio as computer files.

Mix automation

In the mid-1980s, mixing techniques altered significantly with the introduction of *fader automation*, or *flying faders* as they were known. Automation meant that the fader movements from the last pass could be memorized, and perfectly reproduced on the next. The mixer needed only to adjust the faders when a change was required – this meant that a higher number of faders could be controlled by one pair of hands.

Digital audio workstations

Around the same time, dedicated audio editing systems appeared on the market. These computer-based *digital audio workstations* (DAWs) could play several tracks simultaneously in sync with picture. Early models included the Synclavier and the Audiofile. Audio was digitally recorded onto a small internal *hard drive* – unlike today, storage space was at a premium. These systems enabled the sound editor to manipulate perhaps eight or 16 tracks of audio in a *non-destructive* way. Most were made to

Figure 1.4 Current version of the Audiofile SC digital audio workstation (courtesy of AMS Neve).

work in a way that (ex) film sound editors could relate to, borrowing from film technology in their design. The track layout resembled a cue sheet, which moved in the same direction of travel as a Steenbeck. Edit points were located by scrubbing, and film language was used to name functions: cut, fade, dissolve, reel, etc. For a long time, DAWs were used in conjunction with film and multitrack systems, but as stability and reliability improved in the early 1990s, digital sound editing became universally accepted. Early systems were hampered by lack of disk space and computer processing power, despite the fact that audio needs relatively little storage space compared to that required by moving pictures. However, the development of the DAW – and its counterpart, the digital mixing console – meant that audio could remain in the digital domain right up to transmission or the making of the print for theatrical release. The analogue problems of generational loss and noise were at least minimized up to this point.

A number of systems quickly gained industry acceptance: in addition to AMS, DAW systems were manufactured by Digital Audio Research, Fairlight, Digidesign, Akai and SSL, amongst others. Systems in common use included Audiofile, Soundstation, Audiovision, Pro Tools, SSL Screensound, Sadie and Waveframe. As the cost of disk space dropped, and processing speeds increased, manufacturers began to offer systems that could play higher numbers of tracks, and could store an increasing amount of audio.

A few years later, less expensive systems became available that used a standard desktop Mac or PC as an operating platform for the editing software. This led to the development of *platform-independent plug-ins*, which offered the editor additional editing tools and processing devices in the form of software, and which could be used instead of the rack-mounted hardware equivalents. The development of on-board mixing consoles meant that sound could now be edited, processed and mixed within a single system.

Digital video formats

The first digital videotape format, D1, was developed by Sony and released in 1988. This format offered uncompressed digital picture, and four independent channels of 16-bit/48 kHz audio. This was followed by D2 and D3, and then D5 and Digital Betacam – both of which offered four channels of audio at an improved resolution of 20-bit/48 kHz. DigiBeta was intended to replace the Beta SP format, and has become an industry standard format often used for acquisition and mastering. Sony have since introduced a second digital format called Beta SX, which is aimed specifically at ENG users. It uses a higher rate of compression than DigiBeta, and is backwards compatible with Beta – all Beta formats use ½-inch tape.

Since 1990, a number of formats were developed that used compact Hi 8 tapes; however, these have largely been superseded by the digital video or DV format. For professional use, a number of manufacturers developed DVCPRO, DVCPRO 50 with two additional digital channels, and DVCPRO HD, a high definition version. The Sony equivalent is DVCAM. All use extremely compact ¼-inch tape, and record both sound and picture digitally using a miniaturized camcorder. DV is now used as an acquisition format in news, current affairs and documentary because of its low cost and extreme portability.

Non-linear editing systems

Non-linear editing systems (NLEs) became generally available with the introduction of the Avid Media Composer and the Lightworks system in the late 1980s. The reproduction of the moving image required a larger amount of RAM, a faster CPU and more memory than that needed by the DAW. Early systems could only store picture information at low-quality resolution, and the systems themselves were very slow compared to current models. Again, the design of the NLE was based on film technology. Material was digitized into bins (rather like film bins), and the picture and audio track layout resembled that of a Steenbeck. The huge advantage of these systems was that material could be cut into several versions and saved, a task which would be difficult and perhaps impossible using traditional video and film techniques. Editors could also try out a variety of visual effects, add graphics and titles, and audition fades and dissolves before committing themselves to a final cut. Gradually, non-linear editing almost completely replaced tape and film editing. As computer technology advanced, the advent of on-line systems meant that some productions could be entirely completed within the system without any significant loss in picture quality.

Avid, Lightworks and D-Vision dominated the market for a number of years using custom-built hardware. However, as in the case of the DAW, a number of low-cost systems were eventually developed in the late 1990s that used adapted PC and Mac desktop computers: these included Adobe Premier, Speed Razor and Final Cut Pro, amongst others.

Dolby Digital

In 1992, Dolby released a six-channel digital projection system which took advantage of the recent developments in digital sound editing and mixing. Dolby Digital 5.1 has six discrete channels – Left, Centre, Right, Left Surround, Right Surround and a low-frequency effects channel, which carries frequencies below 120 Hz. Using AC3 compression, the six tracks are encoded into a data stream, which is printed on the film print. On projection, this data is decoded back into the original six channels and

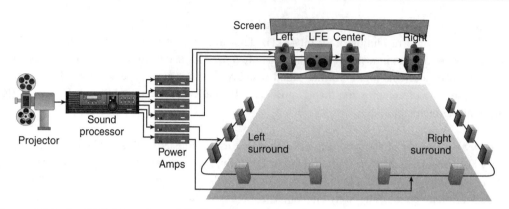

Figure 1.5 Dolby Digital (courtesy of Dolby Laboratories Inc.).

routed through the playback system of the cinema. The use of surround is intended to envelop the audience in the sound field of the film, and enhance their sensory experience through the use of the surrounds and subwoofer. Most digital film mixes and DVDs are currently encoded using this format, although there are other systems such as DTS and SDDS. Dolby Digital can be reproduced in the home using a Dolby Pro Logic decoder.

All film prints using the Dolby system currently carry the digital track between the sprocket holes of the print. The two optical Dolby SR tracks are positioned between the sprockets and the picture. This configuration enables the same print to be used in digital and non-digital cinemas, and is often referred to as a Dolby SRD (Spectral Recording + Digital).

Digital mixing consoles

These were developed in the early 1990s, although this remained a specialized market until the launch of the relatively low-cost Yamaha 02R in 1995. These desks enabled the automation of all mix parameters, including level, panning, equalization (eq) and processing information. Each physical channel

Figure 1.6 Film dubbing theatre with digital console and integrated digital audio workstation (centre) (courtesy of Videosonics/AMS Neve).

on the mixing surface could be used to control a number of inputs, so that very high-end desks could support hundreds of tracks. Since their inception, digital consoles have expanded to cope with an ever increasing track count, and the need for more and more processing power.

Unlike analogue desks, digital desks are not *hard wired* and desk settings can be customized to suit a particular job, saved and recalled at a later date. A mix may also be 'held' on the automation only, without the need to lay it back physically. This facility has led to the technique of *virtual mixing*, and changes can now be made right up to the last minute without the need to pick apart a physical mix.

Digital television

Whilst the television production process has been almost entirely digital for a number of years, digital broadcasting has only recently become a viable reality, and implementation has been slow in coming. There are a number of formats in existence, but all offer a much improved picture and quality in comparison to existing analogue TV formats (NTSC, SECAM and PAL). Standard definition television (SDTV) offers a superior picture and sound quality in comparison with analogue formats, but the highest quality system is high definition television (HDTV). This system uses the highest number of horizontal lines and the highest number of pixels per line to create one frame of high resolution picture – there are three agreed formats:

- 720p – 1280 horizontal lines × 720 pixels progressive.
- 1080i – 1920 horizontal lines × 1080 pixels interlaced.
- 1080p – 1920 horizontal lines × 1080 pixels progressive.

The 16:9 aspect ratio mirrors that of cinema, and the sound format is digital stereo or Dolby Digital (AC3). The improved resolution of digital TV requires a higher bandwidth than conventional analogue TV. MPEG-2 encoding and compression is used to accommodate the digital signal within existing analogue bandwidths.

Digital acquisition

Digital video formats such as DigiBeta rapidly made analogue tape formats obsolete, and whilst this development has resulted in improved signal quality, the tape-based production workflow is more or less the same. The next major step forward was the development of high definition (HD) formats, which support a much higher picture resolution (see 'Digital television' section). Most HD formats support four channels of digital audio at a minimum resolution of 16-bit/48 kHz. Mastering formats such as Panasonic's D5 support eight audio tracks of 24-bit/48 kHz, which are designed for use with multichannel audio (e.g. 5.1 and 6.1 mixes).

In 1999 Sony and Panasonic introduced the 24p format, which was developed to emulate the resolution of 35-mm film. 24p cameras record picture in HD (HiDef) resolution as digital data files on a ½-inch HD cassette. The camera itself is switchable between 24, 25 and 30 frames per second, as well as between 50 and 60 Hz. The format is supported in post production by 24p VTRs and editing systems such as Avid DS Nitris, which can import 24p material, compressed or uncompressed in SD or HD format.

Many television productions which would have originally been shot on 35 mm and digitally post produced (such as high-end drama and commercials) are now being shot using 24p throughout the process, and some film directors have chosen to use 24p for feature-film shoots. The 'film look' is achieved by the use of progressive (the 'p' in 24p) scanning, which results in a sharper image than interlaced scanning. The 24p camera can also be under- or over-cranked just like a film camera to achieve variable-speed effects. Where portability is important, the DVC HD format offers the advantages of DV whilst maintaining the high quality of the format.

Until recently, sound location recording for film and television has been carried out using digital tape-based formats such as DAT. Multichannel recorders that can record audio as sound data files are now replacing this format. These files can be imported directly into desktop computers and editing systems as a file transfer, rather than a real-time transfer. The recorders themselves support a much higher audio resolution than has been previously possible, with recorders typically sampling up to 24-bit/192 kHz. File-based field recorders have been used in a specialized way since 1999, when the Zaxcom Deva was introduced, but in the last year a number of new systems have come onto the market, and the manufacturing of DAT field recorders has ceased.

Digital cinema

Whilst the majority of cinemas now project soundtracks digitally, film itself is still ... film! Whilst it is technically possible to project both sound and picture from a digital system, the high cost of installing the system itself has proved prohibitive to cinema owners and very few cinemas have been equipped with digital projectors. Films can be distributed by satellite or on CD-ROM and loaded into the projection system's hard drive. The main advantage of this technology over print-based distribution methods is that it eliminates handling damage and wear and tear of the print due to repeat screenings. However, some film studios are concerned that the risk of piracy increases when films are delivered for projection on a digital format rather than on film stock.

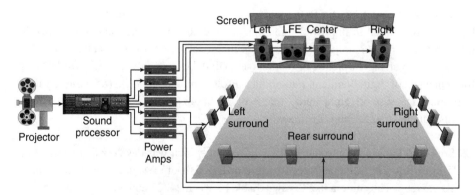

Figure 1.7 Dolby Digital EX (courtesy of Dolby Laboratories Inc.).

Sound surround systems, meanwhile, are still developing. Dolby Digital Surround EX and DTS-ES add a third surround channel (a rear channel) to the existing 5.1 set-up: this is also known as 6.1. Formats such as 10.1 also exist, but are currently only used in computer game technology and domestic digital applications.

Where we are now: post production today

Despite the advances in digital technology, post production is still largely a linear process. Partly, this is due to the way in which post production equipment has evolved. Originally, DAWs were developed by companies that traditionally made other sound equipment, such as mixing desks and outboard gear. Picture editing systems, on the other hand, were generally developed by software companies, and were designed with fairly simple sound editing capabilities – they were not developed in conjunction *with* audio workstations (Avid and Audiovision were the exception). Manufacturers could not agree on a universal operating system, and so all designed their own proprietary software. The end result of this is that the post production 'chain' comprises of a string of stand-alone systems whose only common language is SMPTE timecode!

Project exchange

System compatibility and the exchange of media between systems is one of the most problematic areas of post production, and a number of software programs have been developed over the years to transfer media from one system to another, with varying success. The next major step forward in this area will be the Advanced Authoring Format (AAF), which is still in development. This project exchange protocol is intended to integrate video, graphics, audio and midi information into a single file that can be moved from system to system during post production. AAF is complemented by the MFX format (Material Exchange Format), a file format that enables the transfer of a complete programme or sequence as a self-contained data file.

Merging technologies

Recently developed picture and sound editing systems have much more in common than their predecessors, as most use a standard Mac or PC as an operating platform, rather than custom-built hardware. Some picture editing systems are now available with enhanced sound editing capabilities. This raises the future possibility that a production could be completely post produced within a single system, from picture edit to final mix.

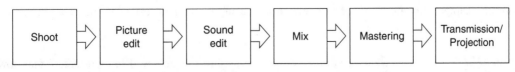

Figure 1.8 A typical 'linear' production workflow.

Filesharing and networking

When picture and sound are imported into an editing system as computer files, these files are saved to *local* storage devices – usually an array of external hard drives. The limitation of local storage is that the media is only accessible by a single workstation, and the drives have to be physically moved, or backed up to a separate device in order to transfer media to any other system. Many productions require data to be shared amongst a number of workstations simultaneously. To facilitate this, *networking systems* exist that can store large amounts of compressed or uncompressed media as data. The media is saved to a multiple array of drives, which are configured to act as a single drive in conjunction with a *server*. A number of terminals can be attached to the server, and multiple users can access the data simultaneously through high-speed data links. In practice, this means that editing, effects, graphics and audio post can take place at the same time, which may be a crucial advantage in fast turnaround projects, which are broadcast immediately after completion.

In a newsroom, for example, all ENG footage is digitized into the system *once*, and everyone can instantly access the same media from the server. Each journalist is equipped with a desktop editing system on which sequences are cut. Voice-over is recorded directly into the system at the journalist's desk, using a lip mic to reduce background noise. If the cut sequence needs to be tracklaid and mixed, it can be instantly accessed by a sound mixer. Once mixed, it can be broadcast directly from the server, eliminating the need for tape handling and playouts.

On a feature film, several editing systems may be configured around a filesharing network. The same media could be used by the assistant syncing up rushes, the assistant doing playouts for the director, and the editor(s) cutting the picture. High-budget films employ several sound editors who each work on a specific part of the tracklay (one editor may lay all the transport effects, another may specialize in laying atmospheres). In this case, each editor has access to all the sound fx being used in the project, and reels can be moved around between workstations until the sound edit is complete.

Storage Area Networks (SANs) use Fibre Channel technology to connect devices up to six miles apart, and can transfer data at very high speeds. This type of network changes the production workflow considerably, as material can be *randomly accessed* at any stage, resulting in a *non-linear* production process.

The idea of a tapeless production process has been around for a long time. Theoretically it is now possible to shoot, post produce and broadcast using file-based technology from start to finish. In practice though, such technology is currently used in conjunction with existing video, film and audio systems. Figure 1.9 shows how acquisition, post production and distribution workflows might be integrated around an SAN. Although SANs are expensive to install, many post production facilities are currently using them, and they will become an increasingly common feature in the post production environment.

In the next few chapters we look at how system compatibility is achieved in more detail, and how project data is exchanged between systems in practice.

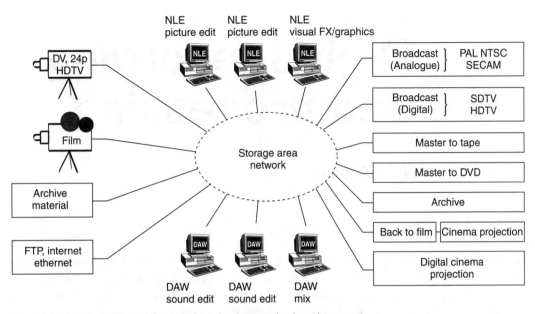

Figure 1.9 Production workflow options in a networked environment.

2 | Digital recording and processing

Tim Amyes

The digitizing and recording of sound and video is a sophisticated and powerful modern technology, but its origins go back to mechanical calculating machines. These work on the simple principle that any calculation is possible using just two digits, 'zero' and 'one'. This technology has enabled the transformation of cumbersome mechanical calculators into lightning-fast electronic calculators and computers, all driven by 'chips'.

The power and speed of the electronic silicon chip has grown exponentially since the 1970s, but it is only comparatively recently that it has developed enough storage capacity to successfully serve the video and audio post production industries.

The whole digital revolution in communications, both audio and video, is dependent on devices called analogue-to-digital converters (often abbreviated as A/D converters or ADCs). These are devices designed to take in an analogue signal one end and to deliver a digital version at the other. They work on the principle of breaking down a waveform into two distinct parts, one part recording the waveform's position in time as it passes a point and the other recording the waveform's amplitude – how loud it is at that point. These time and amplitude readings are repeated and recorded very regularly and very accurately thousands of times a second. The time element, the speed with which the sound is sampled, is called *sampling*. The recording of amplitude is called *quantization*.

Digital signals take one of two possible values when recorded: on or off. These become equally saturated points of recorded data – one positive and one negative. Noise within the system can be completely ignored providing, of course, that it doesn't interfere with the two values recorded. The fundamental unit of digital information is the 'bit' (binary digit), whose state is traditionally represented by mathematicians as '1' and '0', or by engineers as 'Hi' and 'Lo'. The two states can be physically represented in a wide variety of ways, from voltage peaks in wires to pits in plastic discs (e.g. DVDs). Digital media can be transferred from electrical wires to magnetic tape, to plastic CD, and back again with perfect fidelity, so long as the same sampling rate is used throughout. If the speed is only fractionally out, the system will not work properly – there will be synchronization 'timing errors' and possibly no signal reproduced at all.

So what is the advantage of using all this complicated technology just to record audio? Recorded digital sound (and vision) is distortion free – unless overloaded, when severe clipping will occur, 'cleaner' and noiseless. However, its main advantage over analogue sound is that digital audio can be copied

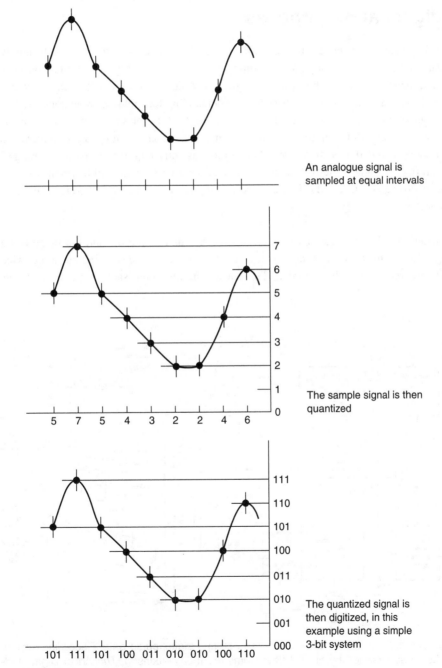

An analogue signal is sampled at equal intervals

The sample signal is then quantized

The quantized signal is then digitized, in this example using a simple 3-bit system

Figure 2.1 The principle of digital audio recording.

many times with almost no degradation of the original signal. It is this that makes it so ideal for audio post production systems.

The digital audio process

In audio, digital sampling above '40 kHz' with quantization above '16 bits' gives high-fidelity recordings, equal to and better than analogue recordings. Providing the analogue-to-digital conversion to the recorder and then the returning digital-to-analogue conversion is of high quality, the reproduction will be near perfect. But audio digital processing is a demanding technology. In analogue recording, the frequency range of the audio equipment need only reach 20 000 cycles a second, for this is the highest frequency we can hear. But in digital recording the analogue signal is immediately converted into a fast stream of code, recorded as a stream of offs and ons, instantly requiring a frequency range 30 times greater than our analogue system. In fact, digital audio requires a bandwidth similar to that required for video recordings – indeed, the first commercial digital audio recordings were made on modified video recorders.

Before processing, the analogue audio is cleaned of signals that are beyond the range of the digital recording system. This is because the rate at which the signal is to be sampled has to be twice as high as the highest frequency to be recorded. This means the minimum sampling rate has to be at least

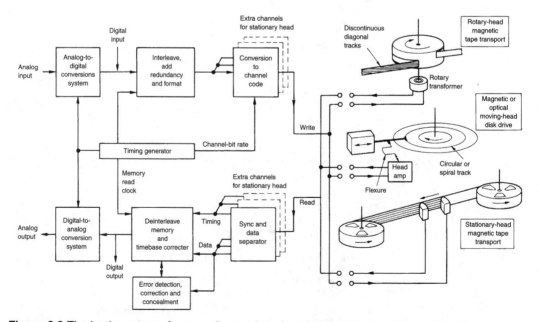

Figure 2.2 The basic systems for recording and storing digital audio data. Recording systems using tape are being superseded by hard disk systems (from J. Watkinson, *The Art of Digital Video*, with permission).

40 kHz, otherwise there could be interference between the digital signal and the audio signal. Such processing is known as anti-aliasing filtering.

The incoming signal is now ready to be *sampled* (in kHz) at regular intervals, and the magnitude determined. The samples are then 'quantified' to give each a quantity value from a set of spaced values (bit values). This means the original signal is now represented by a string of numbers.

Sampling rate and pulse code modulation

The sampling rate is normally greater than twice the highest frequency to be recorded and it essentially determines the frequency response of a system. For successful audio recording, the sampling rate must exceed 40 kHz. Audio CDs, for example, have a sampling rate of 44.1 kHz. The sampling must take place at precise regular intervals, controlled by an accurate quartz crystal clock, similar to the one found in watches. This is the fundamental principle of pulse code modulation (PCM), where the audio waveform is sampled in regular steps. The distance between the steps is dependent on the speed of the sampling. As the sampling rate increases, so does the number of samples. The more samples there are, the higher the quality. However, this also means that the amount of data recorded increases, which will require an increase in storage capacity. Inconsistencies in the sampling, such as timing errors, will induce sound glitches and drop-outs in the reproduced sound. This becomes particularly critical if two or more sounds are being combined together.

When choosing a working sampling rate for a project, it is necessary to take into account the local studio standard and the end application. Music recording studios use the 44.1 kHz CD standard. However, in broadcasting, 48 kHz is the standard sampling rate. In actuality, most people will not notice any quality difference if a sampling rate is above 48 kHz or not, although 96 kHz is now supported on some professional audio systems.

Quantizing level

Quantization helps determine the dynamic range of a recording. Digital audio is recorded in what is known as linear pulse code modulation (LPCM). In this system, there is a 6 dB reduction in noise as each bit is added to the word length. In practice, a 16-bit recording is capable of a dynamic range of around 90 dB. It is restricted by 'noise' or dither, which is added to the recordings to remove the complete silence that 'falls away' when no signal whatsoever is being recorded (the effect is similar to using a noise gate, as described in Chapter 16). The minimum quantizing rate for professional results is 16 bits. In practice, at 20 bits, noise level is below audibility. Although many digital audio workstations and mixing desks support 24-bit recording, they employ longer bit words internally than they offer to the external world. This allows a system to mix many channels of sound together without an apparent reduction in dynamic range. Although the summing of digital channels does increase total noise, it will not be apparent in the audio at the output of the device. In any project the word length should be at least the same or better than that of the final transmitted or delivered material. This ensures that noise is not added during the mixing process. Conversion from one bit rate to another is possible without quality loss.

Storing digital audio data

Unfortunately, memories in a computer go blank when the system is switched off. Since all programs 'evaporate' when switched off, it is necessary to keep a permanent safe copy of them. These copies are recorded in circular tracks on the oxide-coated surface of a spinning disk called the 'hard disk'. In digital audio the use of hard disks is integral to most systems, such as audio workstations, studio multitrack recorders, some field recorders and servers.

The data is recorded or *written* to the actual disk or *platter* in concentric rings by a read/write head fixed to a very light, stiff arm which can swing accurately and quickly to any track, often in less than 10 milliseconds. This gives 'random access' to any bit of data on the disk. Since the disk may well be spinning under the read-head at 5000 revolutions per minute, the data can be read off very quickly. The disk is not scuffed by the head at speed, since it surfs at a micron or so above the surface of the disk on a 'boundary layer of air'. By stacking double-sided disks with ganged pick-up arms, vast amounts of data can be stored and played out from one hard drive. But in order to reproduce the audio correctly, the signals need to be clean and uncorrupted. 'Error correction' systems are used to ensure this. These may, for example, record the same data two or three times on the magnetic disk. Hopefully errors can be minimized or even completely eliminated, allowing the well-designed system to produce a 150th generation sounding exactly like the first.

Tape is sometimes used for storage of digital audio. It shares the same magnetic properties as computer hard disks. Both benefit from developments in heads, oxide and data coding technology. Tape is a recognized backup format in information technology systems and it will continue to be used for both data and video recording, in step with computer technology. Formats that use digital audio tape formats include DAT and digital video. However, the use of tape is now being superseded by systems which record to a variety of fixed and removable drives.

Sample rate (kHz)	MB per track-minute
22.05	2.52
32.00	3.66
44.10	5.05
48.00	5.49

Figure 2.3 Disk space required per minute of track storage (mono) at different sampling rates.

Compression

Digital signals produce a large quantity of data, sometimes too much data for the recording system to cope with. Here, special techniques can be used to compress the data into a smaller recording space. These techniques are regularly used in video recording but are not so necessary in less data hungry audio systems.

In 'companding', the signal is compressed when digitized and recorded. It is then expanded and restored to full bandwidth when played out. In the 'differential encoding system' a recording is made only when there is a difference between each successive sample so that the overall number of recorded samples is reduced.

Compression systems can reduce audio quality, but a well-designed system will produce excellent results. Certain distinct individual sounds such as solo harps may tend to point up any problems. The ATRAC system (Adaptive Transform Acoustic Coding) from Sony uses the principle of 'psycho-acoustic masking' which, like many others, takes into account the fact that some sounds we hear are obscured by other sounds, so we do not hear them. The system reduces the amount of data recorded by selectively ignoring these sounds. ATRAC data reduction is used in the MiniDisc format and SDDS (Sony Dynamic Digital Sound), a digital recording system which can compress multichannel theatrical mixes down to a ratio of about 5:1 compared with the original recording standard linear PCM recording. Dolby's equivalent system is known as AC3, a format which also uses 'temporal or psychoacoustic masking' to compress a signal by a ratio of about 12:1 in comparison with the original recording. This system compresses a multichannel signal down to two tracks and is used to encode audio on a variety of formats, such as DVD, Digital TV and Dolby Digital film prints.

In order to manage the storage and retrieval of data, and maintain the smooth running of the system, digital systems store data changes in a temporary storage space known as a buffer.

Buffers

When digital data is saved to a hard drive, it is saved wherever there is free space on the disk, and is often recorded out of sequence order. This means that when the data is required to be played out, it has to be retrieved and reconstituted by the system in the original correct order. The audio arrives in the *buffer* in *bursts* of data. The buffer or *cache* (also referred to as random access memory – RAM) receives and delays the signal bursts for a short time whilst they are sorted out into the correct order. Once this has happened, the audio may be played out smoothly, without interruption. The buffer works rather like a water reservoir – it must never become empty or the sound will cease to flow. The digital signal processors (DSPs) try to ensure this never happens. The amount of time a system takes to complete this operation is known as latency and is measured in milliseconds.

A single disk drive can access several different data tracks at once, making it possible to have many tracks playing off a single drive at the same time. Since there is no fixed physical relationship between the data tracks, it is easy to move and manipulate audio around the system. Problems can be solved by reducing the amount of work that the disk head is asked to do by merging some of the operations, reducing the number of tracks or using a lower sampling rate. Replacing the hard drive with one which has a faster *access time* can be a solution. The practical ramifications of drive speeds and access times are examined in more detail in Chapter 10.

The buffer helps maintain speed synchronization with other pieces of equipment, ensuring at the same time that the digital data stream of audio is still correctly delivered. It allows other activities in the audio workstation too – scrubbing sound (moving backwards and forwards across a small piece of sound), slipping sound, instant start and varispeed. In addition, buffers can help to dramatically correct corrupt digital data; in fact, sound unusable on one machine may play successfully on another. Buffer activity can be displayed in workstations, indicating how much digital audio data is in the internal buffer memory at any instant.

Digital sound is:

- of almost perfect quality;
- available very cheaply in domestic formats;
- generally able to be accessed quickly, provided it is not in tape format;
- subject to very unacceptable distortion if it reaches any overload point;
- subject to slow run-up times in playout of audio if the system processors are fully loaded;
- without a truly universal format;
- often unable to be confidence checked in playback while recording to ensure that the recording is present;
- liable to a complete loss of the recording if a disk fault develops.

Interconnecting between digital audio systems

At the heart of any audio post production suite is a configuration of components that constitute a *digital audio workstation*. Within its own domain it records and manipulates sounds as digital data. To maintain the highest possible audio quality on a project, all the audio material should be originated digitally. The digital devices need to be of better or equal quality than the workstation itself. For professional purposes, this usually means working at a minimum of 16-bit quantization and 44.1 kHz sampling. To interface digital recorders successfully to a workstation requires some thought.

- The machines need compatible connections.
- They must be set to digital as opposed to A/D (analogue-to-digital) conversion.
- They must be set to the same sampling rate.
- They must be set to the same quantization resolution (or set to conversion).
- A check must be made to see if a synchronization signal is necessary between devices. Timing synchronization in digital audio data streams needs to be highly accurate, for even infinitesimal speed variations may cause problems.

In most audio-post production facilities the video and sound machines will be speed locked via a central high-quality sync generator, which will maintain perfect synchronization between all attached devices. This generator runs at a standard video frame rate and is a timing source that controls the speeds of tape transports – to ensure they all run at the same speed. It may also provide control of a data clock signal known as a *word clock*. This provides a timing pulse to tell the receiving machine the exact start of the data word (hence *word* clock) and how fast it is being transmitted, thus locking the two separate digital data streams together. When only two pieces of equipment are used to make a digital copy, a word clock is not generally needed – for the interconnecting cable, in standard form (e.g. AES format – see below) will have the timing information embedded in the data stream itself. However, when a device like a digital mixing desk is introduced, all the devices used need to be connected to a word clock to keep the system stable. Badly timed audio will suffer from clicks and plops, and instead of silence there may be a regular beating sound. There are various standards for digital audio interconnections. These interfaces use their own particular type of plugs and connectors, which helps, in some way, to reduce confusion.

Digital audio connectors

In the early 1970s, manufacturers began to provide their own proprietary digital audio interface standards, enabling digital audio equipment to transmit digital data without recourse to analogue. Unfortunately, they all adopted their own standards. However, an international standard was developed defining voltage level, data format and connector type.

The international AES/EBU interface is commonly known as AES 3. It supports 24-bit signals; two channels can be carried. It uses the standard type of XLR connector with twin screened microphone cables. Synchronizing word clock information is included in the data stream.

S/PDIF, the Sony/Philips Digital Interface Format, is the two-channel consumer version of the AES/EBU interface. RCA phono sockets are used with standard coaxial cable. It is found on many professional products and is a most popular interface.

The following formats use a separate word clock generator. The connector is usually of the BNC video type (which is also used as a timecode in/out connector).

TDIF (Tascam Digital InterFace) was originally developed for the Tascam DA88, a small modular multichannel machine with eight-track capability. The eight-channel interface uses a 25-pin type D connector and is found on similar multitrack machines of other manufacturers.

The ADAT (Alesis Digital Audio Tape) interface is found on Alesis modular multitrack recorders and others. It uses standard Toslink optical connectors and cables to transmit eight channels.

MADI (AES 10) – the Multichannel Audio Digital Interface – is designed to connect a multitrack recorder to mixing consoles, carrying 56 channels of audio on a single high-grade coaxial cable.

Unfortunately, despite this variety of plugs and sockets, not all equipment is compatible. Sample rate and format converters can be expensive and the quality depends on the price. If it is impossible to transfer sound in the digital domain between two pieces of audio equipment, the audio analogue outputs and inputs of the machines are usually available. Using analogue connections, it is important to record at the correct levels to ensure over-recording doesn't take place – especially if the signal is to be converted to digital eventually.

Professional equipment offers very high quality digital-to-analogue converters and one transfer in the analogue domain will produce hardly any perceptible change. But the more the signal is removed in and out of the digital domain, the more problems there will be. Recording onto an analogue format will bring a certain loss in quality.

Computer files

Audio held in a computer can be transferred to another machine via:

- an 'audio lead' (e.g. analogue or digital AES/EBU 3) in real time;
- a computer interface at, perhaps, high speed (e.g. Firewire, SCSI, USB, etc.);
- a removable medium (e.g. IDE hard drive, Jazz discs).

To record audio as digital data, specific file formats have been developed, and the following are now in universal use. One of the most common file types is the PC Windows-based Wave or .WAV (and its broadcast equivalent .BWAV), which can support any bit or sample rate. Mac-based computers use the Audio Interchange File Format (.AIF), which again supports all bit and sample rates. Digidesign's file format is Sound Designer 2 (.snd), used in their Pro Tools systems.

Some files only contain digital audio; some also contain important edit data relating to the project. In Chapter 4, we look more closely at the various systems and formats used for the transfer of workstation audio data.

<table>
<tr><td>

3

</td><td>

Synchronizing and controlling audio post production equipment

</td></tr>
</table>

Tim Amyes

Audio post production is designed to bring together the various sounds that will make up a polished finished production, allowing them to be accurately edited and mixed together into the final audio track.

The sounds may be music, effects or dialogue, but all must run in *synchronization* with the picture, and do so repeatedly. The sound editor must be able to lay the soundtracks, and the mixer must be able to rehearse and refine his sound mix to perfection. A synchronizing *code* is needed to achieve this degree of accurate synchronization between the various audio sources and the picture. Film uses *sprocket holes*, but on electronic recording systems (video and audio) a standard electronic code is used – known as *timecode*. On computer workstations, where sound is edited and mixed, manufacturers use their own proprietary timecoding systems within the workstations themselves. The timecode may not, in fact, be recorded as a standard timecode signal, but as a reference code to cue files to play when needed. Here the timing information needs to be very accurate – more than ordinary timecode – to allow critical locking to digital signals. But whatever system is used internally in a workstation, the timecode system that talks to the outside world will conform to internationally agreed standards.

Timecode was first used in videotape editing, in order to achieve accurate cuts. A system was needed which enabled each individual videotape picture to be identified or labelled at the right point. Using timecode, it is possible to identify a frame to be edited and perform a precise picture edit exactly as commanded. All these operations are controlled by the hidden timecode system in the workstation.

Sound can be identified by frames too. If a sound source is recorded in sync with picture, each frame of picture will relate to a particular point in time on the soundtrack. The picture frame and the sound frame may, for example, be 1/24th of a second long (running, therefore, at a speed of 24 frames per second – fps).

So, in order to achieve accurate sync, the picture and sound must match each other on a frame-for-frame basis. However, picture and sound can drift in and out of synchronization, even across an individual frame, and this can happen imperceptibly with no apparent audible or visual problem. Dialogue is most susceptible to noticeable sync problems: any dialogue a frame or more 'out of sync' will be apparent, and the *lipsync* will start to look 'rubbery'.

SMPTE/EBU timecode

Timecode is recorded using digital code. It can be equally useful in identifying an audio point on a tape or a hard disk. The code is accurate to at least one video frame in identifying a cue point for synchronization. The time span of a video frame depends on the speed of the system in frames per second. There can be 24–30 video frames in one second. Once the code has counted frames, it then counts seconds, and so on, identifying up to 24 hours of frames. The code then starts again. Each frame, therefore, has its own individual second, minute and hour identification.

Each timecode signal is divided into 80 *pulses* or *bits*, and each of these bits has only one or two states, on or off, referred to respectively as 'one' or 'zero'. The basic signal is a continuous stream of zeros – the signal that occurs as timecode crosses midnight.

However, if the signal changes its polarity halfway through a 'clock interval', the message represents a 'one'. This is a method of modulation known as *biphase mark encoding*. If listened to, it sounds like a machine-gun varying in speed as the time progresses. Distinct changes can be heard as minutes and hours change.

Each frame, therefore, consists of 80 parts. The speed of recording these bits will depend on the number of frames per second multiplied by 80. Thus, in the easy to understand PAL system, there are 25 frames each second multiplied by 80, equalling 2000 clock rates per second. In the American system, there are up to 30 (among others) frames per second multiplied by 80, giving 2400 clock periods per second. This rate of bit recording is known as the clock rate or bit rate. These coded signals are used to represent the numbers recording clock time – the timecode.

To view a timecode display, it is necessary to have eight separate digits. The largest number being recorded is 23 hours, 59 minutes, 59 seconds and 29 frames (USA). To reach this specific time, all the numbers from 0 to 9 appear at some point, and have to be recorded. However, the code to record them is only in the form of zeros and ones.

To record numbers the code is divided into groups of four, each of the bits having only an 'on' or an 'off' state. Bit 1 of the group of four is designated numerical value 1 when switched on; Bit 2 is given the value 2; Bit 3 is given the value 4; Bit 4 is given the value 8 when switched on.

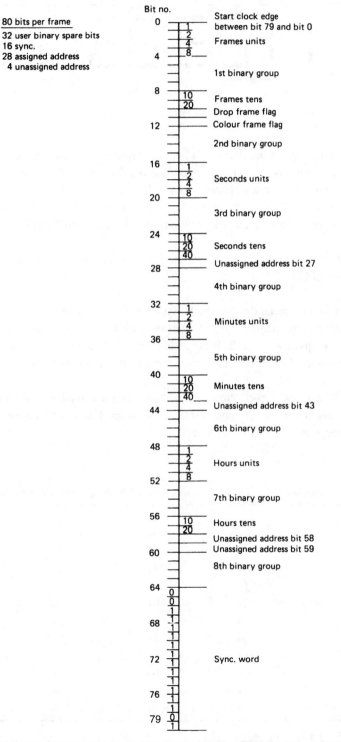

Bit no.

80 bits per frame

32 user binary spare bits
16 sync.
28 assigned address
 4 unassigned address

0 — Start clock edge
between bit 79 and bit 0

1
2
4   Frames units
8

4

1st binary group

8

10
20   Frames tens

Drop frame flag

12   Colour frame flag

2nd binary group

16

1
2
4   Seconds units
8

20

3rd binary group

24

10
20   Seconds tens
40

28   Unassigned address bit 27

4th binary group

32

1
2
4   Minutes units
8

36

5th binary group

40

10
20   Minutes tens
40

44   Unassigned address bit 43

6th binary group

48

1
2
4   Hours units
8

52

7th binary group

56

10   Hours tens
20

Unassigned address bit 58
60   Unassigned address bit 59

8th binary group

64
0
0
1
1
68   1
1
1
1
72   1   Sync. word
1
1
1
76   1
1
0
79   1

Figure 3.1 The basic timecode structure as recommended by the EBU Technical Standard N12-1986 (courtesy of the EBU).

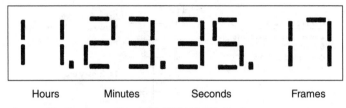

Hours   Minutes   Seconds   Frames

Figure 3.2 Timecode read in the form of an HH:MM:SS:FF display.

Figure 3.3 shows the state of the Bits 1 2, 4 and 8 to represent the appropriate number. For example, to represent value 6: Bit 1 is off; Bit 2 is on; Bit 4 is on; Bit 8 is off. Bits 2 and 4, which are switched on, add up to the numerical value 6. If the number 8 is required, only Bit 4 representing 8 is, of course, switched on. Obviously, if only the numbers 1 and 2 are required – such as units of hours on the clock – a group of only 2 bits is needed, not 4 bits, which extends to 8 and beyond.

This means that, using a group of 4 bits, any decimal number from 0 to 9 can be coded (in fact, the maximum number is 15).

The time information is spread through the 80-bit code: 0–3 is assigned to unit frames; 8–9 to tens of frames; 16–19 to units of seconds; 24–26 to tens of seconds, etc. Interspaced between the time information are eight groups of 4 bits each. These 26 bits are called *user bits* and to access their code special equipment is necessary. This information cannot be added once the code is recorded.

User bits record additional information such as video cassette number, the scene and take number, the identity of a camera in a multi-camera set-up, and so on. The user bit information can be changed by re-entering new user bit settings as required.

Decimal number	*Group of four consecutive bits, value when switched 'on'*			
	1	2	4	8
0=	0+	0+	0+	0
1=	1+	0+	0+	0
2=	0+	1+	0+	0
3=	1+	1+	0+	0
4=	0+	0+	1+	0
5=	1+	0+	1+	0
6=	0+	1+	1+	0
7=	1+	1+	1+	0
8=	0+	0+	0+	1
9=	1+	0+	0+	1
(15=	1+	1+	1+	1)

Figure 3.3 Chart of a 4-bit code.

Other user bits are available to give standard information. Bit 11 gives colour recording information to prevent flashes on edits, Bit 27 is a field marker or phase correction bit and Bit 43 a binary group flag. Bit 10 gives specific timecode-type information. Bits 64–79 are always the same in every frame, since they tell the timecode encoder when the end of frame is reached and also whether the code is running forwards or backwards.

Timecode and speed

Six different frame speeds are used in timecode. It is preferable to use a single timecode standard throughout a production. However, it is sometimes necessary to use different timecode standards within a production – for example, where film (running at film speed) is edited using equipment running at video speeds. Lack of clarity on this issue often causes a number of problems in post production. Practical methods of handling timecode and differing frame rates are discussed in detail in Chapter 11.

Film speed (24 fps)

Twenty-four frames per second timecode is universally used in film for theatrical release. It is also used in high definition formats.

PAL and SECAM (25 fps)

In Europe the television system operates at exactly 25 frames per second; this speed is used in the production, distribution and editing of video material. This conforms to the European Broadcasting Union's PAL and SECAM systems. The speed is historically based on the use of a 50 cycle per second alternating current mains generating system. Television film is shot at 25 fps and transferred to video at 25 fps, the standard telecine speed. Films for theatrical release produced in Europe are normally shot at 24 fps, but this creates problems when the material has to be transferred to a PAL-based workstation for editing. There are a number of ways in which this speed change can be handled within the production workflow. However, the final release print will be at the standard 24 fps.

Films shot at 24 fps are usually shown on television at 25 fps this reduces the actual running time by 4 per cent and raises audio pitch, which can be electronically pitch shifted down.

NTSC (30 fps)

Originally this was used with NTSC black and white television. At its inception there were problems when transferring film to video. Film was shot at 24 fps and the system ran at 30 fps. The two are incompatible. To overcome this problem, additional frames are added as the film is transferred, periodically holding the film in the projector gate for three instead of two television fields (3:2 pull-down). This tends to give film a slightly jerky look. To produce better motion rendition film can be shot at 30 fps, which results in a 1:1 film-to-video frame ratio. On transfer to video, the film and separate sound have to be slowed down slightly (0.1 per cent) to accommodate the NTSC video system.

This system is used for high definition television, television commercials and music videos, but adds 25 per cent to costs. The system is not relevant outside NTSC countries.

29.97 non-drop frame code (NTSC)

In America there is a 60-cycle generating system and this has been tied to a nominal frame rate of film at 24 frames per second. This speed was standardized when the sound film was introduced in 1928 (the slowest speed possible for apparent continuous motion). In most non-video environments, this SMPTE timecode is related to the 60-cycle line frequency combined with a 30 frame per second code rate, as with domestic televisions.

29.97 drop frame code (NTSC)

With the development of the NTSC colour systems, it became necessary to find ways to make the new colour broadcasts compatible with the existing black and white sets. The monochrome signal had to remain the same, so an additional colour signal was added after the frame had been sent, simply by slowing down the video frame rate to 29.97. Thus, clock timecode and the TV signal run at slightly different times. This means using a 30 fps timecode system will result in a drift against an NTSC video recording, which is 0.1 per cent lower than the nominal 30 frames per second speed. In fact, the code gradually falls behind actual real time. The code is therefore modified – to maintain speed, numbers have to be dropped out of the code. Over a period of 3600 seconds, i.e. one hour, there is a loss of 3.6 seconds (0.1 per cent). This is an error of 108 frames, and so 108 frames have to be dropped, through the hour, if real time is to be retained. In fact, actual frames of video or film are *not* lost, only frame *numbers* are skipped to keep clock time. To identify drop frame timecode, user Bit 10 is activated. On identifying this, the timecode discards two frames every minute – except for every tenth minute; $60 \times 2 = 120$ frames per hour are admitted, except 0, 10, 20, 30, 40 and 50; $120 - 12 = 108$ frames per hour are therefore dropped. Code-reading devices will usually flag up which type of code is being read.

SMPTE drop frame timecode therefore matches real time (within two frames a day). The lack of certain frames in the code can create further problems in operations. A synchronizer may be asked to park at a time that does not exist! However, most synchronizers will deal automatically with this.

30 drop frame code (NTSC)

This finds limited use in film origination in the USA to improve motion rendition when the requirements are for television use.

Identification and labelling

Timecode systems can seem even more tortuous when timecode frame rate and picture frame rate do not match. Film can be shot at 24 fps and may have 30 ps timecode on the audio. Whatever the code chosen, continuity throughout the project is essential. In Europe there are fewer problems, but if material is shipped to the USA it must be well labelled (and vice versa) if there are not to be problems. The identification of the type of timecode is most important on material if systems are to synchronize in post production.

Longitudinal timecode (LTC)

In order to record timecode, only a limited audio band width is needed – something between 100 Hz and 10 kHz. This compares with a frequency response of 40 Hz to 18 kHz for high-quality audio recording. No special arrangements are usually made to record this 'analogue' signal, which consists of a series of on and off pulses.

Simple analogue systems used to record longitudinal timecode can find it difficult to reproduce square waves, and strong, interfering frequencies can be generated. This means that timecode recording levels are carefully chosen to minimize possible bleed into adjacent tracks, or crosstalk. Crosstalk can be reduced by recording the code at a lower level. However, if the level is decreased too much, the signal will itself be susceptible to crosstalk from adjacent tracks, and from the background noise of the tape itself. Typical recording levels for timecode are between −6 and −15 with respect to the zero or line-up level. These levels are preset. A typical timecode waveform is shown in Figure 3.4.

LTC can be recovered using ordinary analogue amplifiers at speeds down to one-tenth of normal playing speed, or 2.5 frames per second. Unfortunately, timecode readings become unreliable as tape speed approaches zero; audio output is proportional to speed, so as speed reduces to zero, output reduces to zero. These problems are overcome in digital workstations, where timecode is recorded as part of the overall timing of the system and timecode can be accurate down to a part of a frame.

Video recorders provide continuous timecode from the LTC head and offer the facility to change to *vertical interval timecode* or VITC (recorded within the picture lines) at low speed. Operating in automatic, they swap between these sources as required, using LTC as the correct code and VITC for timing. The correct switch settings should be checked on a machine before use.

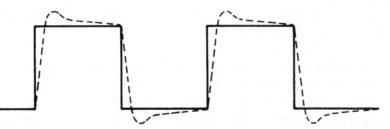

Figure 3.4 A typical timecode waveform superimposed over the ideal theoretical waveform. The overshoot at the edges and the small amount of tilt are caused by phase response errors.

LTC in digital equipment

In a digital audio workstation, longitudinal timecode will lock primarily to the relevant audio sampling frequency being used – this is modified to timecode speed. The digital signals need to be very accurately synchronized and timed if the data is to be correctly analysed and reproduced. If digital audio is to be combined successfully in any way with other digital audio, the signals will be synchronized to a common source. Even if the signals are out by 10 parts per million, audio problems can result.

In digital tape formats, timecode is recorded within the device on its own dedicated tracks. This can provide timecode at all speeds either directly or through a pseudo source.

Timecode is in itself a digital signal, but it may not be recorded in its basic digital form. Although the in and out plugs of a recorder may be labelled as longitudinal timecode, within the recorder timecode may well be more sophisticated. This is the case with DAT. Although appearing to record LTC, it uses the subcodes of the audio data to create perfect looking timecode of the type required. Care must be taken when recording timecode on tape formats, as it is possible to record an external timecode that is not synchronous with the sampling rate. This causes problems in digital post production, as there is a conflict between the need to lock to the digital audio reference and to slave lock to the timecode. Because the two sources are different, one will fail. Either the digital system clicks or you slowly lose sync.

Using LTC in practice

Timecode can be recorded:

- At the same time as the audio material is recorded.
- Before the audio material is recorded by 'pre-striping' the tape. Audio and/or video is then recorded without erasing the timecode – this is now rarely necessary.

When recording on a video format, timecode must start at the beginning of each picture frame to ensure frame-accurate editing. This is achieved by referencing the code generator to the video output.

All synchronizing devices need time to lock up, but there are particular problems with tape transports. These need 3–5 seconds of code or *pre-roll* before the required start point in the programme material, to enable one device to synchronize successfully to other machines. In reality, it is sensible to record 30 or more seconds of pre-roll code before a recording starts.

It is often recommended to start a programme's code at 10.00.00.00, i.e. 10 hours, rather than 00.00.00.00 – this avoids any problems with the machines crossing *midnight* timecode, at which point they become confused. In fact, it is wise to start code at 09.58.00.00 – 9 hours 58 minutes, allowing time for announcements, tone and colour bars. It is also easy to read the duration of a project; you just ignore the '1' in the hours' 'ten' column. This may seem trivial, but under pressure it makes for easy operations.

Operationally, timecode should not be looped though devices by using the timecode in and out sockets, for if one fails so will the rest. It should be divided through a simple audio distribution amplifier and fed individually to each machine.

Recorded longitudinal timecode signals suffer from being copied; however, most audio and video recorders automatically correct timecode problems by *regenerating* the code, and timecode can be transferred between machines without difficulty. It is important to confirm that longitudinal code is being regenerated by checking the machines' settings and the internal menu. Code may stand one direct copy but this is not desirable. When copying timecode using a stand-alone timecode regenerator, both machines and the regenerator need to be locked to a common sync generator – this could come from the sync output of one of the machines, whether it be a digital audio or video recorder.

A number of methods of regenerating the timecode waveform are possible, depending on the severity of damage. However, built-in timecode readers are very reliable and problems are few.

Advantages of longitudinal timecode

- LTC is always available if a tape has been pre-striped.
- LTC can be read at high speeds.
- LTC is suitable for both video and audio recording.
- LTC is a standardized system of delivering timecode.

Disadvantages of longitudinal timecode

- LTC cannot be accurately read in stationary mode, or read whilst jogging at slow speeds. To overcome this, timecode is often visually superimposed or sometimes *burnt into* a recorded videotape picture. This displays an accurate visual reference, even in stationary mode.
- LTC must occupy a dedicated track.
- LTC is subject to severe degradation when copied – but regeneration automatically compensates for this.

Vertical interval timecode (VITC)

Vertical timecode is applicable to video recorders. It provides correct timecode information when the video machine is stationary and is more useful in audio post production because of this.

Vertical interval timecode is similar to longitudinal timecode but does not occupy a dedicated track. Instead, it is inserted into two unseen lines of the video picture, using a pulse format which doesn't need to be of the biphase type (as with longitudinal timecode), since time information already exists within the videotape recorder's own system. The modulation system used is known as a non-return to zero or NRZ type.

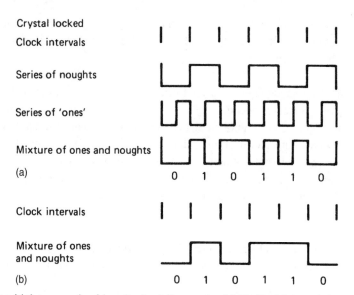

Crystal locked
Clock intervals

Series of noughts

Series of 'ones'

Mixture of ones and noughts

(a)

0   1   0   1   1   0

Clock intervals

Mixture of ones
and noughts

(b)

0   1   0   1   1   0

Figure 3.5 (a) The biphase mark of longitudinal timecode. (b) Vertical interval timecode in the video system is level responsive – a certain level represents peak white, while another level represents black. The black level between the frames is used to represent zero and the higher level nearer the white represents one.

For practical reasons, a vertical interval timecode word has to be recorded twice. This prevents any loss of signal through 'drop-out' on the magnetic tape. If minute faults occur in the magnetic oxide of the tape during manufacturing, or through poor physical handling, a momentary loss of signal or a *drop-out* may occur. Videotape recorders are able to detect picture drop-outs and can replace a faulty line with part of a previous one, stopping a visible flash occurring on the screen. This error concealment process is hardly noticeable. It can be even used to replace damaged frames of film transferred to videotape. However imperceptible this system may be to the eye, it is not a technique that can be used on faulty timecode. Replacing a timecode word with the preceding timecode produces obvious inaccuracies. Instead, timecode is recorded within the picture frame more than once, as a safety precaution.

In the SMPTE system, the vertical interval timecode signal is inserted not earlier than picture line 10, and not later than line 20. It is sited in the same positions in both fields. The EBU gives no specific position for VITC words, but recommend it should not be earlier than line 6 and not later than line 22. International specifications for timecode also cover the use of timecode in digital video recording (where an 8-bit representation of the code is used, recorded on lines 14 and 277) and in high definition television systems for the interchange of material.

Both vertical interval timecode and longitudinal timecode will be encountered in audio post production using video systems.

Advantages of vertical interval timecode

● VITC is available whenever the picture video is available, even in stationary mode.
● VITC does not require any special amplification – if the VTR can produce a good television picture, it will reproduce VITC.
● VITC does not take up what might be valuable audio track space on a VTR.
● VITC is applicable to any video recording system, provided it can record up to 3.5 MHz.

Disadvantages of vertical interval timecode

● VITC cannot be read at very high tape spooling speeds. The individual design of the video recorder affects its ability to read timecode at high speed. It is not normally possible to read VITC at more than double the speed of the videotape. As the code is part of the video signal, VITC fails as the picture breaks up in spool mode.
● To be used successfully in audio post production, VITC has to be recorded via a transfer from the original videotape.

Burnt-in timecode

Burnt-in timecode is not actual timecode as such, but a superimposition (in the picture frame) of the code which is recorded on a videotape. As it is literally burnt in to the picture frame, it cannot be removed once the recording has taken place. It is only really useful when handing over a tape to be used in a system that does not support timecode, because accurate notes can still be made using the timecode references that appear in the visual display on the screen. However, it cannot be used for any *machine control* purposes. Burnt-in timecode is sometimes mistakenly referred to as BITC.

It is often useful to have the current frame timecode visually displayed in the monitor of a system during post production. Most professional video decks have the facility to superimpose timecode, and this may be switched on or off, as well as moved around inside the frame. For most applications, this is more satisfactory than actually burning code permanently into the picture, where it may actually cover up an important picture detail.

MIDI timecode (MTC)

MIDI stands for Musical Instrument Digital Interface – a system which is designed to connect various electronic instruments together. It can be likened to the roll on a mechanical pianola – instructions are only given on what notes to play. No recording of the actual sounds are made – it is simply a very sophisticated switching system with many applications. MIDI is well suited to personal computers, and using the correct software, MIDI outputs can be sourced through the computer's own printer and modem ports. Being a serial bus system, all the information and data is transmitted on a single cable looped through to the various pieces of equipment. Anything transmitted by a source or master goes to all the receivers, each of which is set internally to a specific channel. Very large MIDI systems are likely to be run by their own controller. Musicians can use the system to lock together various synthesizers, which operate rather like a multitrack tape recorder, so tracks can be 'recorded' and more sounds built up. The interface is fully digital and works on binary codes that specify type, destination and the operational speed of the system. Music-based digital audio workstations use MIDI interfaces that can offer audio post production capability by locking to a video player. In order to use longitudinal timecode, MIDI has to be transformed through an interface into data which is compatible.

MIDI timecode contains two types of message. The first updates regularly and the second updates for high-speed spooling. In fact, one frame of timecode contains too much data for MIDI and so it takes two frames to send a frame of information – this then needs to be resolved to achieve frame-for-frame synchronization. In MIDI timecode operations, synchronous time information is provided through a sequencer. Sequencers are devices that allow the recording of a sequence of notes or switching information for a number of notes that might make up a song. Sequencers need to work on a defined time scale. The sequencer acts as a slave and is fed with MIDI timecode derived via an interface from the master timecode machine (SMPTE/MTC). The sequencer can then slave any appropriate MIDI device.

Workstations themselves offer MIDI in and out connections to allow control over other MIDI devices such as musical sequencers – this is necessary if workstations are to be used for music recording. Most equipment is likely to have MIDI in, MIDI out and MIDI through connections, and be capable of transmitting information on 16 separate channels. MIDI through is used to feed more than two MIDI devices; MIDI out is only used at the beginning of a bus. MIDI connections, unlike other systems, are standardized by five-pin DIN connectors. The maximum length of a MIDI cable is about 15 metres. The MIDI interface can either act in master mode, where it generates pseudo timecode, or in slave mode, where MIDI commands are derived from the SMPTE/EBU timecode readings.

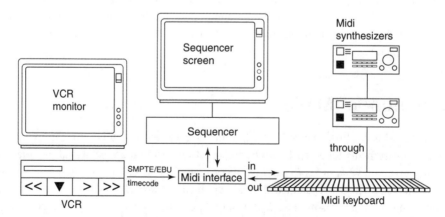

Figure 3.6 A MIDI system set-up for music composition using a video recorder with timecode. The sequencer synchronizes the timecode information.

Controlling equipment through synchronization

Audio post production equipment must be synchronized with other pieces of equipment for successful operation. Many of these operations will happen automatically and once set up need not be reconfigured. Types of equipment which can be synchronized together using timecode include:

- Workstations.
- Video recorders.
- DAT machines.
- Modular digital multitrack recorders.
- MIDI musical equipment.
- Hard disk video and audio recorders.
- Telecine machines.

Synchronization modes

At the heart of a synchronizing system is the master. The actual running speed of the master is controlled either internally or externally – it is normally locked to a word clock or video reference and produces timecode which the slave machines must follow or *chase*. This master may be a workstation or possibly a video recorder.

Master

In a simple audio post production system, a video recorder (using tape or hard disk) and a digital audio workstation (without a picture facility) are locked by timecode. The video may well be the master and is operated either manually from its own controls, or remotely from the workstation controls. Videotape recorders are more difficult to control from external sources and can be slow to lock up if they are in slave mode. Their mechanical transports take time to run up into synchronization as the various servos come into play. Digital workstations without these mechanical restrictions can follow almost instantly. Only the size of their data buffers and their processing speed will affect their ability to synchronize quickly. Audio workstations can usually act in either slave or master mode. Workstations with the ability to record in picture as a digital file avoid the need to wait for an external video machine to lock up, which makes them very popular with editors, who can often operate their systems at very high speeds.

Slave synchronization

A machine in slave mode is one forced to run in synchronization with the master, which supplies the master code. Equipment that may be slaved includes workstations, DAT machines, modular digital multitracks, video recorders, workstations, telecine machines, etc.

Digital systems do have problems in maintaining slave synchronization because the speed has to be maintained without digital glitches or clicks. As a slave, a workstation can lock up to speeds varying

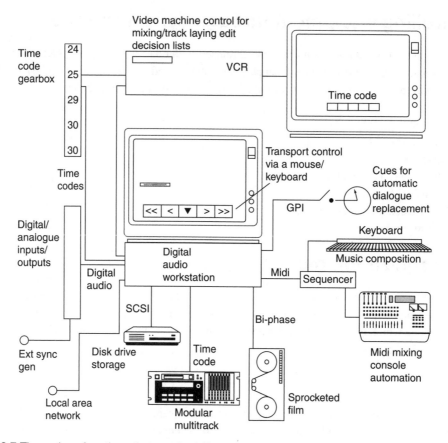

Figure 3.7 The various functions that can be offered by an audio workstation.

from 10 per cent off the mean, without inducing audio problems – much depends on its computer processing power. Some are able to operate in reverse in lock, others cannot. The more sophisticated workstations allow remote control of the external master, such as a video recorder, from their own built-in controls.

Synchronous resolution defines the limits to which the system will normally synchronize. These parameters can usually be optimized for the particular equipment used, often a videotape recorder. A synchronizing slave will not, in fact, synchronize exactly – it will tend to lag behind. This lag can range from one frame to sub-frames. Within workstations the resolution must be highly accurate – any lack of synchronization between the two tracks of a stereo signal will be audible.

Chase synchronizer

The simplest of slave synchronizers is the single chase unit, offered either as a stand-alone unit or as an integral option for a machine. These are the most common form of synchronizer. They are available for many machines at a cost of little more than a good microphone. In operation, the machine senses the direction and speed of the timecode fed into it and follows or chases the master wherever it goes, comparing the slave's speed to match that of the master. Only timecode is needed to control the machine, giving speed and location information. Chase synchronizers are used when conforming and transferring sound to and from a workstation, using an edit decision list, and also for synchronizing a separate soundtrack with picture in telecine operations.

Trigger synchronization

In slave synchronization the slave follows the master up to speed and then carries on to control it at synchronous speed. This is unnecessary with many types of equipment that are capable of running at synchronous speed within themselves. In this situation, a device triggers or switches over to its own speed control system when synchronous speed is reached. This is used in some audio workstations, but it is sometimes switchable. The incoming signal can then be locked to the workstation's own clock reference, which may be more accurate and may produce higher audio quality. However, trigger

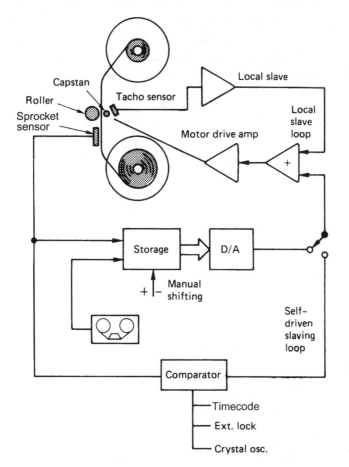

Figure 3.8 A telecine servo control system.

synchronization can be unreliable, as the speed may drift on one of the systems. If a system is only trigger locked, it is also unlikely to notice if timecode has jumped and is no longer continuous. This can happen through an edit – since the system is no longer reading code, it may wrongly assume that the original code is still there and unbroken. If, however, the device follows and chases timecode it will detect a timecode break when dropping out of play or record, or when a warning is flagged up. A chase and trigger lock will warn of a break, but will still freewheel over the problem, usually for up to about 4 seconds – this figure can be user-defined in some manufacturers' equipment.

The control of tape/film transports

To keep transports in synchronization with an external reference or timecode, or with their own internal speed generator, motor control circuits are used to ensure that the speed of the tape or film is exactly in step with the controlling source. Tape machines (video or audio) and telecine machines (which convert film images to video) synchronize through servo control systems. The drive motors receive their synchronizing information from timecode recorded on the tape, or alternatively from a free-running tachometer wheel driven by the sprockets of the film or telecine machine. These codes or pulses are then referred to an incoming master signal. The machine's speed is continuously altered to match the exact speed of the incoming signal. This is called a 'closed loop' servo system, so the precise speed of the transport is slaved to an external source.

4 | Audio transfers and file formats

Hilary Wyatt

The proliferation of DAWs and digital recorders on the market, and the development of both proprietary and non-proprietary file formats, has meant that the act of transferring audio between systems has become one of the most complex areas of post production. The lack of complete agreement on universal standards and the rapid development of new formats means there is often not a clear 'pathway' along which audio can travel through the production chain.

Compression

A signal can be reduced in size, or *compressed*, so that it occupies fewer channels and less space, enabling it to be transferred or distributed more easily. *Lossless* compression techniques ensure that the original signal can be reproduced perfectly once decoded or *decompressed*. *Lossy* compression techniques discard the parts of a signal that are considered redundant, which means that once the signal is decoded it is not a perfect reproduction of the original. The abbreviation *codec* is often used to describe a *c*ompression/*dec*ompression software program or *algorithm*.

Audio compression

Audio data can be compressed for ease of distribution (e.g. MP3 music files downloadable from the Internet). Compression reduces the data rate, resulting in faster transfer times, but also results in a loss of quality that may be unacceptable in some circumstances. For many professional applications, audio is transferred uncompressed. The data rate, and therefore the quality of the transfer, is expressed by the bit depth (which defines dynamic range) and sampling frequency (which defines frequency bandwidth). For example, audio CDs are encoded as PCM 16-bit/44.1 kHz stereo (uncompressed).

Data compression

AC3 and Dolby E are *bit-rate reduction* algorithms used in film and TV that compress multichannel signals into a compact data stream. Both formats use *perceptual coding*, a data reduction format which ignores inaudible frequencies that are masked by other elements in the signal. Once decoded, the human ear cannot perceive the difference between the processed signal and the original. Digital compression can enable up to eight channels of data to be transferred within the same bandwidth needed to transmit one analogue signal.

AC3 is a general-purpose algorithm that is used to compress a signal before transmission. It can be used to reduce a 5.1 multichannel mix into a data stream using the least number of bits possible, without resulting in a corresponding loss of quality. It is most often used to compress a 5.1 mix down to a form that can be physically fitted onto the edge (between the sprockets) of a film print, or encoded onto a DVD, or broadcast in a Digital TV format. Dolby Digital and Pro Logic surround systems use AC3 encoding, and indeed the terms are often used interchangeably.

Dolby E is used to compress multichannel audio purely within the production environment prior to transmission, and is not intended for broadcast. It can compress a multichannel signal (up to eight channels) down to a two-channel data stream, which can then be distributed via digital I/O, such as an AES/EBU pair. Unlike AC3, a Dolby E signal can be encoded and decoded many times without any signal loss, and is encoded in a way which corresponds to the video signal on a frame-for-frame basis.

There are two ways in which audio can be transferred from one medium to another. The first is a *linear* transfer, which may be analogue or digitally recorded. The second is a *non-linear* or *file transfer*, which results in a digital-to-digital, bit-for-bit copy of the original.

Linear transfers

The simplest yet most inflexible way of copying audio is to make a linear transfer onto a format that is universally accepted for certain applications. Audio can be encoded as an analogue signal or, more usually these days, encoded as *streaming* audio digital data. There are many standard uses for this type of transfer. Audio CD is the standard format for sound fx libraries. DAT is still (at the time of writing) the most widely used location recording format, as well as being used to create safety recordings or *snoop* DATs in ADR sessions, stereo mix masters and safety backups of stereo mixes. Tascam DA88, with its eight available tracks, is perfect for playing out the *stems* of a 5.1 mix, with a corresponding stereo mix occupying the remaining two tracks, or for storing stereo splits such as a DME (Dialogue, Music and Effects).

When making a transfer, the main concerns to bear in mind are:

● *Avoiding distortion and noise.* In the case of analogue-to-analogue and analogue-to-digital transfers, extreme levels can result in distortion and clipping. Unlike analogue distortion, digital formats have an upper limit, beyond which the signal is lost – visually this appears as though the top of the waveform has been cut off.

Low levels can result in a poor signal-to-noise ratio, and system noise can be introduced. To maintain a constant-level reference, a 1 kHz line-up tone should be placed at the start of the recording to be used

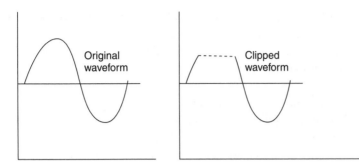

Figure 4.1 Digital 'clipping'.

as a level reference for subsequent transfers. Transfers should be recorded at the European TV standard operating level of −18 dB or the US TV standard operating level of −20 dB, which is also the operating level for film (including European film).

● *Locking to 'house sync'*. In a post production facility where there are a number of video and audio systems working in conjunction with each other, all machines must be 'locked' together to an external centralized sync reference or *house sync*. House sync can take the form of a video signal that is sometimes known as *black burst*. Without this reference, the two machines creating the transfer will gradually drift out of sync with each other. There are some machines, such as CD players, that do not have the capability to lock up in this way.

● *Maintaining sync reference*. If the transfer is to be made with timecode, specify the correct rate. If the delivery format does not support timecode (e.g. CDR), place a *sync plop* 48 film frames before the first frame of picture, and clearly label the cover with this information. This is equivalent to the '-3-' mark on an Academy leader and the '-2-' mark on a Universal leader. Where the editor has placed a leader at the start of picture, check that a corresponding sync plop has been laid, even if timecode is also used. Clicks and drifting sync problems can occur in streaming digital-to-digital transfers if the two machines are not clocked together correctly.

● *Correct sample rate*. The standard delivery sample rate for film and TV has always been 48 kHz, and film production is increasingly conforming to this. However, some facilities have a house rate of 44.1 kHz because most of their sound libraries (based on CD) work at this rate. Ideally the sample rate will be maintained throughout the transfer, but where it is not, there are two options. The first is to take the audio in via the analogue inputs. For example, a sound fx CD that runs at 44.1 kHz can be recorded into a DAW running at 48 kHz via the analogue-to-digital converter (ADC). The second option is to pass the audio through sample rate conversion software. These devices can vary in quality, and a poor program can introduce clicks and distortion into the converted audio. However, most recent digital editing systems can handle sample rate conversion satisfactorily, and some can even convert 'on-the-fly' during playback.

● *Error correction*. Where digital errors have occurred in a recording (perhaps because of bad head alignment), the 'missing' bits can be reconstructed by the error correction facility of the playback machine. This occurs most commonly on professional DAT machines, where an indicator light on the front panel will indicate the amount of error correction taking place. In practice, if the amount of correction is quite high, then a DAT that will not play on one machine may play on another that has a wider 'tolerance'.

The advantage of linear transfer over other methods of exchanging audio is that it results in a simple tangible copy, which is important for some applications such as archiving. Linear transfers are also generally made on formats that have universal compatibility. However, for some applications there are some serious disadvantages to linear transfers:

- All cue idents and identifying information is lost.
- All fade, level and cut information becomes fixed.
- 'Silence' is copied as well as the audio, with no differentiation.
- Handles are lost. (Handles are the audio either side of a source cue that can be peeled out to make a better edit.)

In order to maintain editorial flexibility, there are a number of methods used to transfer audio, which preserve names, handles and other information to a greater or lesser extent.

Linear auto-conforming

This method is mainly used to reproduce the audio edit from the picture editor using the original production sound. This takes place as a real-time transfer from the source machine (typically a VTR or DAT deck) into a DAW. Linear conforming from DVD rushes requires the use of specialist machines such as an Akai DD8 or Tascasm MMR8. Some DAWs (but not all) have internal conform software, including Akai, Fairlight, Sadie, Audiofile and Pro Tools, which uses Digidesign's 'PostConform' software. The sound editor receives the source audio and an edit decision list (EDL) in an appropriate format (CMX, Sony, etc.). The EDL specifies a roll number, plus source in and out points for each sound cue, and its start and end points in the assembly, as well as track mapping information.

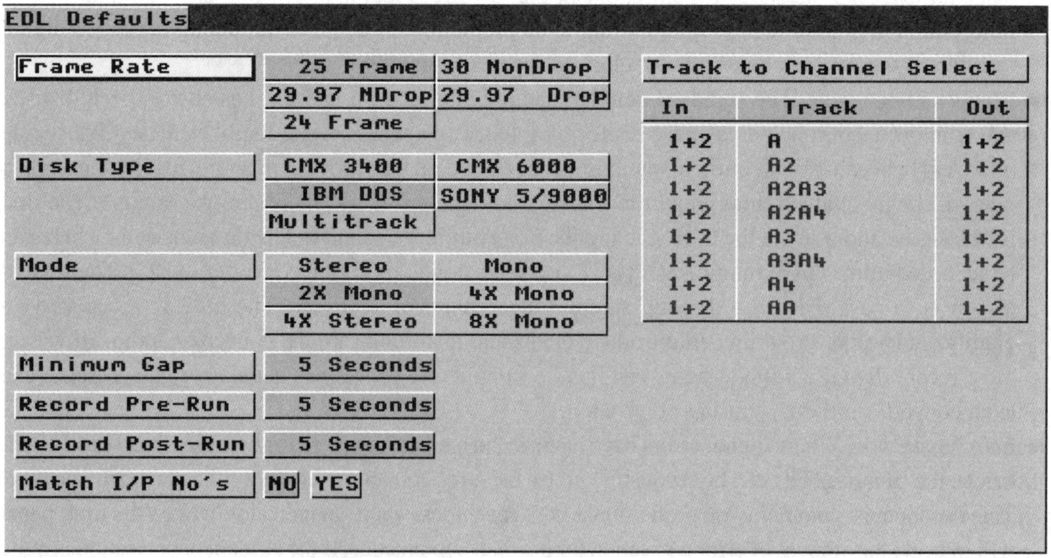

Figure 4.2 Audiofile SC Conform Settings window (courtesy of AMS Neve).

The EDL is loaded into the DAW and the appropriate frame rate and track configuration are specified, as well as the desired handle length. The conform software will prompt the operator to load each source roll. The DAW will control the source machine (via Sony nine-pin) locating the start point of each cue required from that roll, and will automatically drop into record before moving on to the next cue. Once all the cues are recorded in, the conform function assembles the audio into the correct position in the project, thus replicating the picture editor's original edits within the DAW.

The disadvantages of auto-conforming are:

- Only supported by some DAWs.
- Only accurate to −/+ one frame.
- Transfers take place in real time, which is slow in comparison to file transfers.
- There are often problems with EDLs, resulting in audio conforming out of sync.
- Some systems place limitations on the number of tracks that can be contained in an EDL. For example, where only four tracks can be conformed at any one time, it will be necessary to supply separate EDLs for each group, i.e. tracks 1–4, tracks 5–8, and so on. (PostConform handles eight tracks simultaneously.)
- All source material needs a reel ID and timecode reference. Non-timecode sources (e.g. music from CD) will not conform.
- Doesn't carry over fades or level settings. Dissolves need to be converted to cuts before making the EDL.

The advantages of this method are:

- Original source audio is used, rather than audio direct from the picture edit system that may have been digitized incorrectly.
- Only 'used' audio is transferred.
- The sound editor can preview audio quality during transfer.
- EDLs can be read as simple text and edited on a PC.
- Material can be conformed as twin track, mono or a stereo pair as required.
- Inconsistency in source sampling frequencies can be corrected by conforming in through the analogue inputs with negligible effect.
- If the material is re-edited, it can simply be re-conformed in the DAW to the new cut.

File transfers

File transfers are transfers in which audio is exchanged as data within a file format. Whilst the transfer cannot be monitored, the data will be copied at a far higher speed than streaming transfers. For example, hours worth of rushes can be imported into an Avid in a matter of minutes. (Actual transfer rates will be dictated by the bus speed and drive type, as well as sample rates, bit rates and track count.) Where a project is imported as a file transfer, sync should be *sample accurate*. File transfers are already in common use for certain applications, but the recent development of file-based field recorders means that audio can be transferred as files from the location recording to the cutting room and from there to the sound editor's DAW. These recorders support very high bit rates and sample rates – most can record 24-bit/96 kHz audio. This development will significantly affect the entire audio post production workflow, as it will very quickly replace tape-based recording as the accepted way of working. However, as the dust is still settling on this issue, it is too early to say what will become standard procedure in the future.

System compatibility

There are a limited number of manufacturers' file formats that may be read directly by other systems as native files. For example, Aaton Indaw can directly read DVD-RAM disks recorded on a Deva field recorder, and is used for syncing up rushes. Pyramix can read Pro Tools SDII; an Akai can read Fairlight files. However, DAWs are now being designed to offer a high degree of project interchange with other digital equipment on the market using an array of interfaces and file formats. (Nuendo and Pyramix can handle AES 31, OpenTL, OMF, etc.)

Proprietary/non-proprietary file formats

The long-term question mark over the future of file formats is one of ownership. A proprietary format is one wholly owned by a private company, who may choose to change a format or cease to support it altogether. Where projects are archived in such a format, it may become impossible to access the material at some point in the future because subsequent software releases are not backwards compatible or the format has ceased to exist altogether. (Delivery requirements for feature films still specify real-time

WAV	Does not support timecode	WAVE Microsoft audio file	Accepted by most DAWs. Often used to transfer short non-timecode clips via CDR or Internet
AIFF/ AIFF-C	Does not support timecode	Audio Interchange File Format (-Compression)	Mac format used by Avid to export media
SDII	Supports timecode	Sound Designer II, developed by Digidesign	Used by Media Composer, Film Composer and Pro Tools, and other Avid products
BWAV	Supports timecode	Broadcast Wave File, adopted as standard by AES/EBU. Adopted as native file format for AAF	An enhanced WAV file which supports metadata. Mac and PC compatible

Figure 4.3 Audio file formats.

playouts: this sidesteps the issue and 'futureproofs' the mix!) Non-proprietary file formats are ratified by international standards organizations such as the EBU and AES, whose protocols can be implemented by equipment manufacturers without the possibility of future non-compatibility.

File types

There are essentially three types of file format. The first is simply a file that contains a basic unit of audio with no edit information. The second type of file contains not just the audio data, but data about the data or *metadata*. This can include a time stamp (timecode), sample rate, roll, scene and slate number – in fact, most of the information that would otherwise be written on a sound report sheet.

Broadcast WAV (BWAV)

This format has now become the industry standard file format, and is supported by all manufacturers. The file itself is a platform independent *wrapper*, which contains an audio cue and associated metadata stored in *chunks* within a *capturing report*. This consists of the following information:

- *Coding information* – this details the entire transmission history of the cue though the production chain.
- *Quality report* – contains all events that may affect the quality of the audio cue.
- *Cue sheet* – each event is listed with time stamp, type of event priority and event status, as well as location markers.

BWAV files are the standard format for file-based field recorders, and now can be imported into Mac-based edit systems such as Avid. Some DAWs can read BWAVs directly, together with their time stamp.

Monophonic and polyphonic WAV files (BWAV-M/BWAV-P)

With the introduction of multitrack field recorders such as Deva II and Fostex PD6, files can be recorded as polyphonic or *wrapped* multichannel files. These files behave as a mono file, but in fact are four, six or even eight separate channels interleaved into a composite file. It is simpler for some field recorders to write one polyphonic file than several monophonic files of the same size (particularly those which do not write directly to hard disk, but use media with slower RAM speeds, such as DVD-RAM). Polyphonic files transfer more quickly than the equivalent number of mono files. Unfortunately, there are ramifications further down the line, creating yet another compatibility issue. For example, at the time of writing, later versions of Avid will read polyphonic files, but these cannot be imported directly into Pro Tools. Pro Tools users can, however, access all the tracks of a polyphonic file through the use of specialist software such as 'Metaflow'. To extract one or two channels from an interleaved file is quite a complex task for a DAW to achieve in real time during playout, and many current systems will fail to play polyphonic files.

The third type of file can be thought of as a *project exchange* format, which not only contains audio data as basic sound files (e.g. SDII), but also contains all the edit list information required to move an

OMF1 and 2	Open Media Framework. Proprietary format developed by Avid	Transfers audio and picture files. Supported by picture edit systems and most DAW manufacturers
AES 31	Audio Engineering Society 31-3	Non-proprietary format. Simple audio-only interchange format supported by most DAWs (not Pro Tools)
OpenTL	Open Track List. Developed by Tascam	Proprietary open format supported by Nuendo, Pyramix, Logic and others
AAF	Advanced Authoring Format, still in development from Avid, Microsoft and others	Transfers picture and audio files. Expected to supersede OMF in near future, and largely based on OMF1. Open format
MXF	Material Exchange Format. Adapted from AAF	Transfers picture and audio files in a streamable format. Used specifically to transfer completed projects around a networked environment. Open format

Figure 4.4 Project exchange formats.

entire project from one system to another. Some formats are designed to carry over both picture and audio information, whereas others are designed to be used for audio data only. Because of the manufacturing differences between systems, most project exchange formats need to be modified by third-party conversion software before they can be read successfully by the destination system (see below).

OMF1 and 2

OMF was designed by Avid to be an *open* format available to all so that Avid files could be read by other manufacturers' workstations. It is a complex piece of software that carries across both video and audio data. Due to its complexity, the software has been adapted by DAW manufacturers to suit different systems. Because of this, it is important to bear in mind the specific requirements of the target DAW when creating the OMF, as the transfer will fail if these requirements are not met. There is little difference between OMF1 and OMF2 in terms of transferring audio: the latter has enhanced video features, but DAW manufacturers will generally specify which format should be used with their systems. Audio can be *embedded* in the OMF in the shape of WAV, AIFF and SDII files, but the OMF can also work almost as an EDL, or *composition-only* file, which links to the actual media.

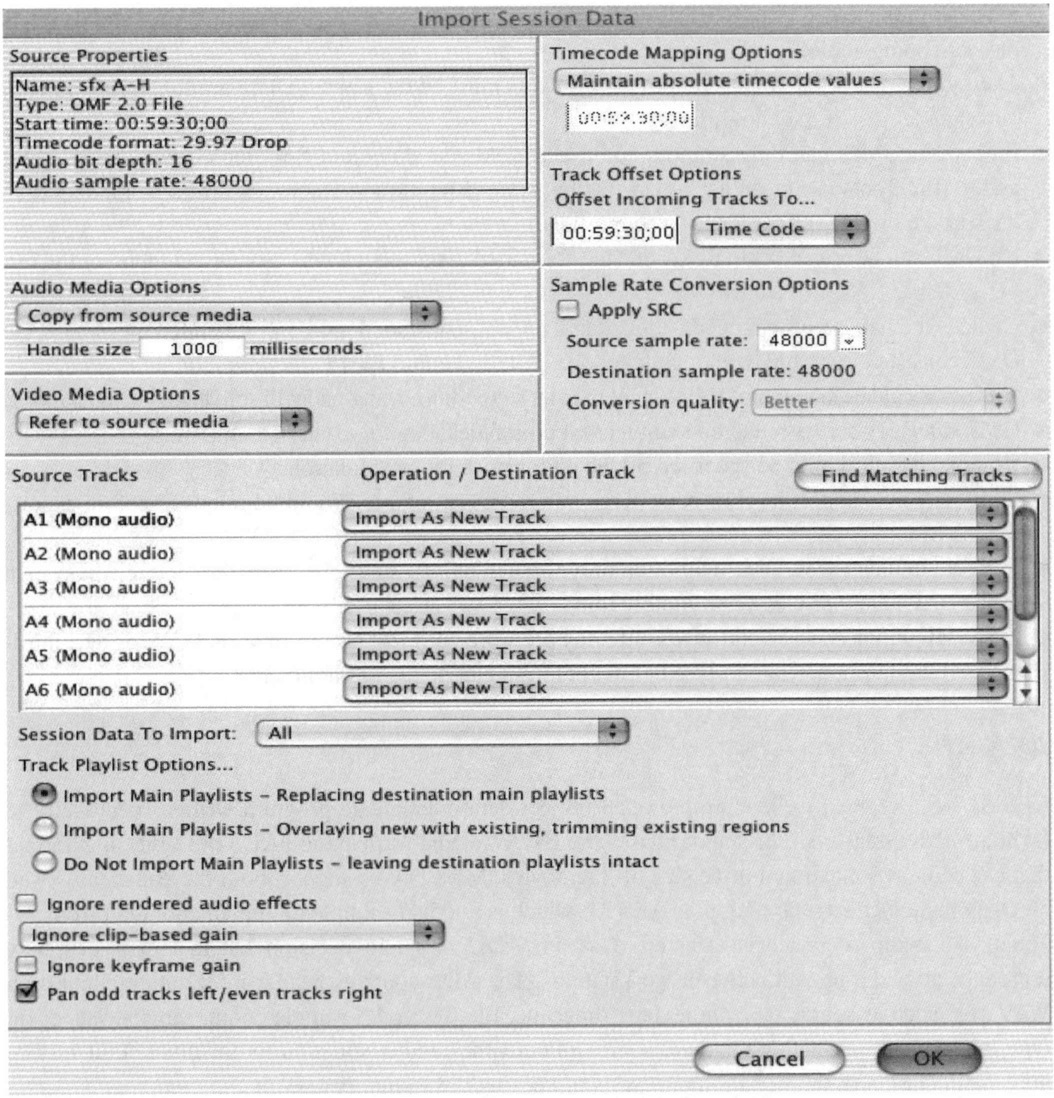

Figure 4.5 Pro Tools: OMF Import Settings window (courtesy of Digidesign).

To transfer audio via OMF, the editor should copy and then consolidate the media onto a new drive, having first deleted the video track. The file should then be exported with the appropriate file format and OMFI composition.

There are some considerations to take into account when transferring via OMF:

● It is preferable to keep the OMF to a manageable size, bearing in mind any limitations of the target system. Most DAWs will read an OMF only if a certain set of parameters is adhered to, and these must be implemented prior to exporting a file (e.g. it might be necessary to remove all audio effects or overlapping audio).

- The sample rate of the import and export systems must be the same. (There are tools that can convert sample rates if they do differ.)
- Embedded OMFs can only support 16-bit audio. Most DAWs and field recorders support 24-bit audio. (Pro Tools users can get round this by using 'Metaflow', which can link a 16-bit sequence back to the original 24-bit location sound files.)
- An OMF can be created onto a PC- or Mac-formatted disk – check which one is suitable for the target system.
- Decide whether the system needs 'embedded' or 'non-embedded' audio. (Embedded means that the OMF consists of one file, which contains the EDL and audio files.)
- All cues should be in mono – OMF cannot transfer stereo-linked cues and will split them into dual mono.
- OMF can carry across rendered symmetrical crossfades, but *not* asymmetrical fades.
- 'Handle' length should be specified by the editor prior to consolidation.
- OMF may not support level or eq settings – in most cases these should be removed before exporting the file.
- In most circumstances, OMF will often only work in one direction, i.e. from the picture edit system to the DAW, and *not* vice versa (unless third-party software is used).
- The OMF will need to be converted either by software built into the destination DAW or by a third-party file translation program (see below) before it can be read as a project.

AES 31

AES 31 was designed to be a simple open project data interchange program written specifically to facilitate audio transfers. The standard uses the BWAV media format and FAT 32 (32-bit) disk format that can address a maximum drive size of 2 terabytes. Most DAWs support both these standards, and all DAW manufacturers have adopted AES 31, which is relatively simple to implement (apart from Pro Tools). An audio sequence is transferred as an 'ADL' or audio decision list that can accurately reproduce an audio project between workstations. The ADL contains header information about each WAV clip, such as source tape name, tape timecode, file timecode, sample count, sample rate, event type, source and destination channel, source in/out times and destination in/out times. Edit history information has been omitted in the interests of keeping the format simple.

Some considerations to bear in mind when transferring via AES 31 are

- This standard is *not* supported by picture edit systems.
- This standard is supported by all DAW manufacturers except Digidesign.
- Some file-based field recorders will be able to deliver audio directly as an AES 31 project (Aaton Cantor X and HHB Portadrive).
- As with OMF, it is still important to consolidate the project before exporting. This guarantees all audio is present prior to transfer.
- The ADL is ASCII based, which means that it is readable in text, not unlike a traditional EDL. This is useful in identifying why a transfer may have failed.

- AES 31 supports multiple channels – for example, mono and stereo clips. However, in some cases the events in the transfer can become jumbled up because the ADL may translate 'mixer settings' that were wrongly carried over from the source DAW.
- AES 31 will describe fades very accurately, including asymmetrical and non-linear fades.
- AES 31 will carry across level changes.
- It will not carry across eq, dynamics and real-time automation parameters at present, although this is intended in the future.
- The AES 31 file will need to be converted either by software built into the destination DAW or by a third-party file translation program (see below) before it can be read as a project.

OpenTL

This is a proprietary open format that was developed by TimeLine and now maintained by Tascam. Like AES 31, it is an audio-only format that carries across fade and level data in a simple text EDL. It is supported by Nuendo, Pyramix, Media Magic and AV Transfer, amongst others.

AAF (Advanced Authoring Format)

This is an initiative by a group of companies (led by Avid and Microsoft) to develop a complete open interchange format. The file format is designed to handle all the data required during the composition of a project. It is intended to encompass video, graphics, text and audio (the *essence*), as well as synchronization and timing information (the *metadata*). It can access material held on a number of devices, and can be used to 'pull' together work in progress from a number of sources during editing. Due to the corporate force behind the format, it is likely to become widespread and is being implemented on picture editing systems such as Avid and Final Cut Pro. It draws heavily on the OMF1 format, which it is expected to fully supersede. As there are a large number of companies involved in its development, it is thought that this format, when implemented, will be a more truly 'open' standard than OMF.

MXF (Material Exchange Format)

This type of file is an open project exchange format primarily intended to move simple completed projects around from server to server for distribution and transmission, or to a data streaming device for archiving purposes. MXF can also be transferred over the Internet via FTP (see below). The file itself is a self-contained *wrapper* that contains all the picture, audio, compression and synchronization metadata associated with a project. MXF is a simplified variant of AAF, and the two formats are designed to be interchangeable via invisible file conversion.

File conversion software

There are a number of software tools that act as a conversion interface between all the industry standard file formats described in this chapter. As such, they are crucial to the successful transfer of audio

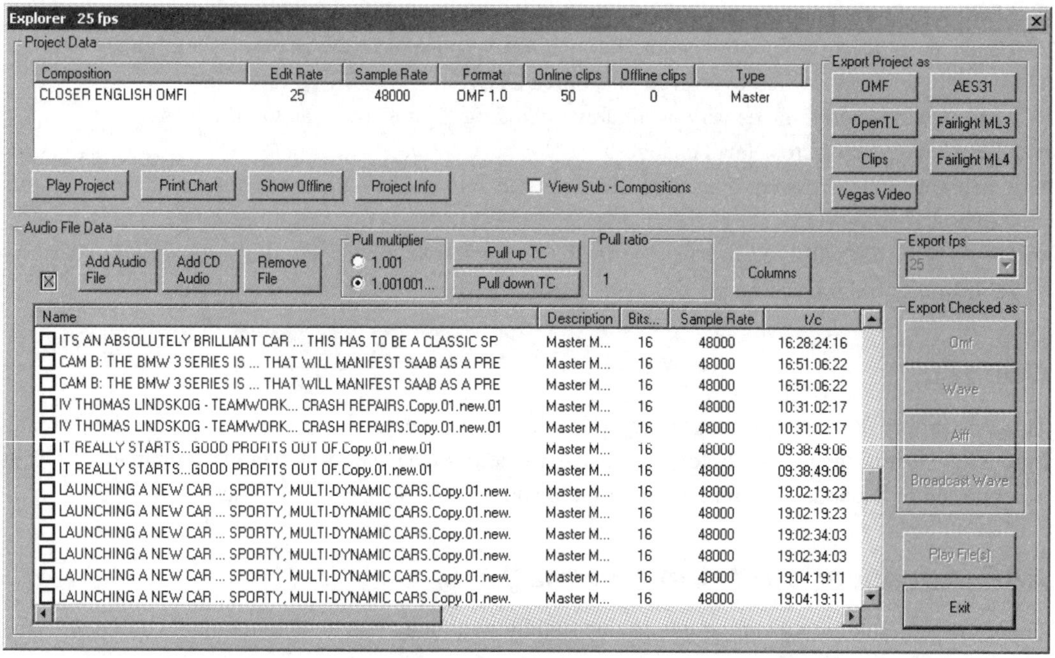

Figure 4.6 AV Transfer: Explorer window.

files and metadata. Typically the user opens a project (in any one of the supported formats) and is taken to the main utility page.

At this stage the user can see the elements that make up the project (individual sound recordings and EDLs). The user can click on any cue or EDL, and view and/or play the audio. The user then exports the project to any one of the supported formats. Each export format has its own properties specific to the destination system. Software programs in common use are:

- AV Transfer. This will enable the import/export of projects between most of the major editing systems. AV Transfer will handle WAV, AIFF, BWAV, AES 31, OMF1 and 2, OpenTL and other formats. It will also convert between different sample rates and bit depths, and can operate in any timecode rate. It will play OMF files directly on any PC and can be used simply to translate OMF files to OMF files where a different 'version' of OMF is required by the target DAW.
- *Media Magic*. This is mainly aimed at Mac users, and will import/export between OMF 1 and 2, AES 31, Akai, Fairlight ML, Pro Tools, Zaxcom Deva, Nagra V and others.
- *Media Toolbox*. AMS conversion software for Audiofile users.

File transfer workflows

It is worth reiterating that the use of file-based recordings in place of tape-based recording is still relatively new and standard workflow patterns have not yet been established. Partly it is clear that this is because there are too many standards in existence, and partly because the level of data a hard disk

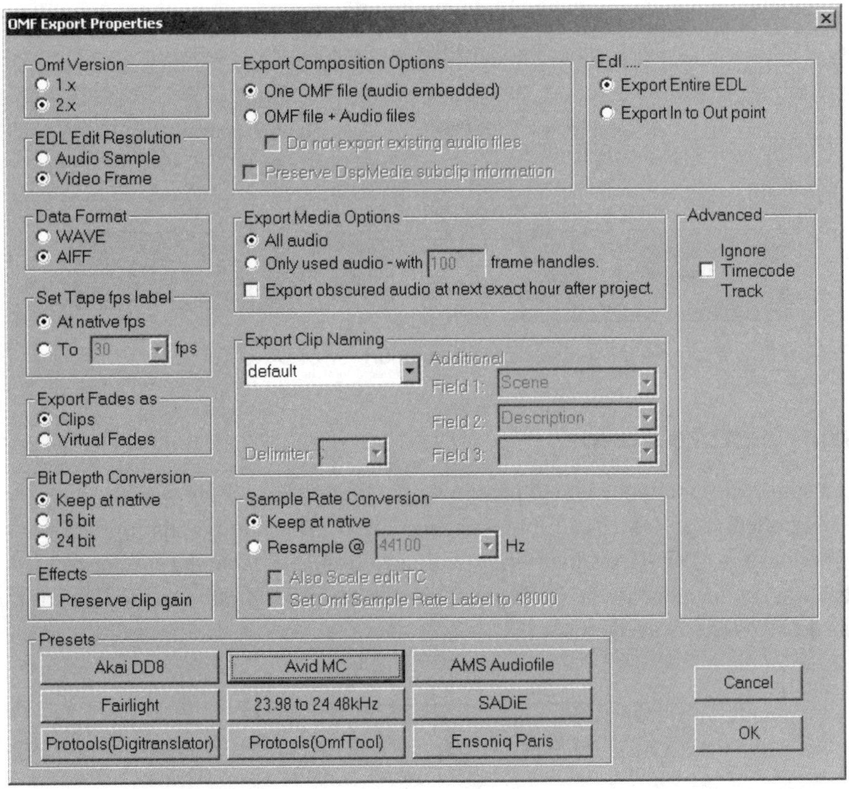

Figure 4.7 AV Transfer: OMF Export window.

recorder can capture is not yet supported all the way down the post production chain. In practice, this often means choosing the easiest way of achieving system compatibility with existing equipment, and this will vary from production to production. However, very broadly speaking, there are two methods of transferring audio between the location recorder and the DAW.

Non-linear conform

The location sound files are synced up to picture by a third party, or within the picture editing system. Once the project is edited, an audio EDL is created in the format appropriate to the target DAW, and if possible, this should be accompanied by a drive containing the media. If a drive is not supplied, *all* audio files must be copied to the hard drive of the DAW, as it is not yet possible to conform BWAV files directly from location DVD-RAMs. (This can actually be quite a slow process because of the read speed of the DVD-RAM.) Once the time-stamped BWAVs have been imported, the files are conformed using the EDL: conforming in this way is very quick. However, few systems can currently do this, and certain rules must be followed (for example, Pro Tools uses 'Titan', which will conform BWAV-M files only). Unless the sound editor is happy to use that particular system, a further export must be made onto the preferred DAW before work can begin. The main advantage of this method is that *all* takes are available to the dialogue editor on-line.

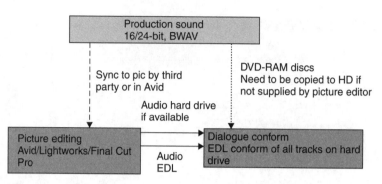

Figure 4.8 Workflow example: non-linear conform (courtesy of Dave Turner).

OMF from picture editor

The location sound files are synced up to picture within the picture editing system. Once the project is edited, it is exported as an OMF, and imported into the DAW using the appropriate file conversion program. The location DVD-RAM/hard drive will need to be available during the edit to access any material that was not included in the OMF, e.g. alternate takes. Using this method means that lengthy file copying sessions are avoided.

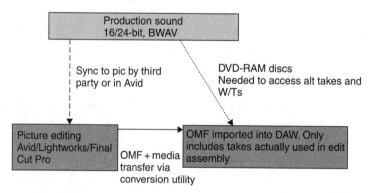

Figure 4.9 Workflow example: OMF from picture editor (courtesy of Dave Turner).

Some preplanning is required to ensure successful transfer of audio between the systems involved in a particular workflow, and it is very important to try to do a test transfer in advance.

Network systems

File formats such as OMF and AES 31 are designed to bridge the gaps between a series of individual systems used in the post production chain. However, where audio and video are saved as files within a system that exists as part of a network, many of the problems of transferring media are overcome, and system compatibility is no longer an issue. Storage Area Networks, for example, can link a high number of different devices and applications by using a common file format and shared media. Such systems are designed to transfer not only picture and audio, but all accompanying metadata as well, and will become increasingly common in filmmaking as well as broadcast operations. Media may be transferred between networked picture editing systems and DAWs in either direction. In the long term, this will replace the need for project exchange software, which often can only operate in one direction, resulting in a rather inflexible workflow. Network configurations are discussed more fully in Chapter 10.

FTP (File Transfer Protocol)

File Transfer Protocol is a platform-independent protocol which can be used to send sound files over the Internet. The most reliable file formats for this purpose are AIFF, WAV, BWAV and MP3. As the system supports more than one platform, it is important that each file has the correct file extension attached to the file name, namely file.aif, file.wav, file.mp3. Text files can also be sent with the file extension file.txt. File formats that cannot be used, particularly Sound Designer 2 files, should either be converted to AIFF or 'stuffed' into a WinZip or Stuffit archive (file.zip, file.sit). It is useful to stuff all files prior to sending, as this reduces file size through lossless compression, resulting in a quicker transfer without a drop in quality. Many facilities have an FTP site on which they might place, for example, ADR from an ISDN session, which can then be downloaded by the client. Sounddogs.com uses an FTP site to deliver fx orders, which can then be downloaded almost instantaneously.

5 | Video, film and pictures

Tim Amyes

In order to fully understand the post production process, a basic knowledge of picture recording and operations is necessary, and in this chapter we look at the technology behind video recording and shooting on film.

Moving images, whether they be stored on film or videotape, are nothing more than a series of stationary pictures, each slightly different from the previous one and each capturing a successive movement in time. When these pictures are shown or projected at speed, the brain interprets them as continuous motion, a phenomenon known as persistence of vision.

It takes approximately 1/20th of a second to visually register a change in an image. When the images change at greater speeds than this, we see pictures that move.

Film

A film camera records pictures photographically. This is achieved by using a claw which engages onto a sprocket and pulls down the film so that it is in a stationary position in front of the lens. A shutter operates for a fraction of a second to expose the image on the film and then, while the shutter is closed, the next frame of unexposed photographic film is positioned (pulled down) for the shutter to open again for exposure. The full camera aperture of a piece of standard (35-mm) film is embraced by engineers under the term 2K. It requires a resolution of 2000 lines (as opposed to the 625 or 525 lines used in analogue television) to emulate the film frame.

The image can be viewed on a film editing machine (usually a Steenbeck), in a telecine machine or on a projector. Projectors are usually equipped with an enclosed xenon lamp as the light source. Four international width formats are used: 16 and 35 mm for television use, and 35 and 70 mm for cinema projection. Seventy-millimetre prints are, in fact, shot using a 65-mm camera negative. Sixteen-millimetre is often exposed, particularly in Europe, using a larger than standard sized aperture called Super 16 mm designed for theatrical release and for 16 by 9 widescreen television. Super 35 mm is a similar format. On both, the soundtrack area is 'incorporated' into the picture area, offering an enlarged picture aperture.

The intermittent movement produced in a film camera at the picture aperture is unsuitable for sound recording, which requires a continuous smooth motion. So sound cannot be recorded in synchronization at the 'picture gate'. Therefore, sound is usually recorded separately on an audio recorder, which runs in sync with the camera; this is known as the *double system*. Cinema release prints are printed with the soundtrack to the side of the picture and the track is replayed some frames away from the picture, where the intermittent motion of the claw can be smoothed out (above the picture-head in 35 mm and below it in 16 mm). Film normally runs at a speed of 24 frames per second (fps) in the USA and 24 (or sometimes 25) fps in Europe. Film is almost always transferred to video for editing. This can create problems, for film shot at 24 fps does not synchronize easily with our video systems – which are based on the local AC power frequencies of 50 Hz (Europe) or 60 Hz (USA).

Telecine

Film editing is usually carried out in digital non-linear editing suites. This means the film has to be transferred to video prior to being loaded into the system. The film-to-video transfer is made using a *telecine* machine. These machines handle the film very carefully and can produce images of the highest possible quality. However, high-quality images are not always necessary if the transfer is only for editing and viewing purposes in the cutting room. In this case, a basic *one-light telecine* transfer is made, saving time and cost. Once editing is complete, the edit is remade using the original camera negative, which is carefully colour corrected or *graded* from shot to shot, so that lighting and colour imbalances can be graded out.

Video

The television system works on a similar principle to that of film, again exploiting persistence of vision. Here, the image is produced as an electrical waveform, so instead of focusing the image onto a photographic emulsion, the television camera focuses its image onto a light-sensitive charged-coupled device (CCD), made up of many thousands of elements or pixels. In a high definition system

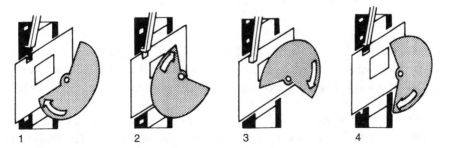

Figure 5.1 The claw and shutter mechanism of a film camera: 1, exposure of film; 2, claw pull-down; 3, film advance; 4, shutter opens gate (courtesy of F. Berstein).

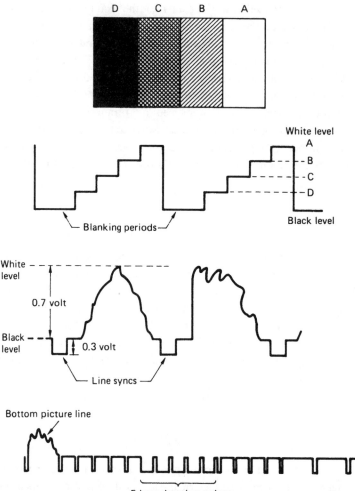

Figure 5.2 The conversion of a picture to a television signal. The picture ABCD is converted into various signal voltages. In the 'blanking' period between the lines, the 'electron beam' flies back. The synchronization pulses are added at the end of each field to keep the camera and receiver in step (courtesy of Robinson and Beard).

there may be as many as 2 073 600 pixels. Each of these picks up variations in light, and breaks the picture down into its separate elements. These are sent to a recorder or to the transmitter for broadcasting. The pictures are, in fact, *interlaced* together every half frame or *field*, to reduce flicker.

To reproduce the picture, an electron beam that is 'controlled' by the camera is directed onto a phosphor screen that forms the front of the television monitor, tracing a picture. The beam is muted during its 'fly-back' journey from the top to the bottom of the screen in order to stop spurious lines being seen. The electron beam scans across the picture at a speed of 15 625 times per second. Each vertical scan is called a 'field'. The European analogue TV standard PAL 625-line system presents 25 complete pictures every second and the screen is scanned 50 times. In America, on the NTSC system, 525 lines are produced at almost 30 frames per second (actually 29.97 frames, a change which became necessary because of the introduction of colour to US black and white television), with scanning taking place 60 times. Since the picture has been split up by scanning, it is important that it is reassembled in step with the camera. Two pulses are sent to ensure this, one a line scan pulse and one a field scan pulse.

High definition television (HDTV) has radically improved quality in television pictures. Using this system, video images can be recorded and distributed to emulate the resolution quality of film. The number of scan lines is increased to between 720 and up to 1250 from the minimum of 525, although there is no agreement on a common standard as yet; 1080/24p is most commonly used in the USA and Europe. The 1080 defines the number of lines and 24p the speed and scanning pattern, which is *progressive* rather than interlaced, resulting in a sharper image. High definition systems create an immense amount of data, which requires a much greater amount of storage space to be available in an editing system, compared to that required by conventional analogue formats. This also means that equipment requires higher data interconnection speeds and processing capacity.

Once there were a few standard video recording formats, now there are more than 10 in current usage. Audio post production suites, serving both the television and film markets, have to be ready to accept many of the current formats used in the field, although the Beta SP format has become somewhat of a standard for off-line worktapes in the industry. It is usually not cost-effective to have all VTR decks permanently available to replay all the formats required (although some are backwards compatible, e.g. a DigiBeta machine will play Beta SP), so these are often hired in as needed. They must be able to interface with equipment already available in terms of audio, picture, synchronization and transport control – so it is therefore important to check that the machines interface with existing equipment already installed. Most professional video recorders follow the Sony P2 (officially called DB-9) protocol for interfacing with editing suite controls using a nine-pin socket. The DV formats, however, tend to use the more modern 'Firewire' computer interface, both for control and exchanging data.

Remote-1 in (9P)/Remote-1 out (9P)		
Pin No.	Controlling Device	Controlled Device
1	Frame Ground	Frame Ground
2	Receive A	Transmit A
3	Transmit B	Receive B
4	Transmit Common	Receive Common
5	Spare	Spare
6	Receive Common	Transmit Common
7	Receive B	Transmit B
8	Transmit A	Receive A
9	Frame Ground	Frame Ground

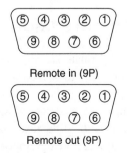

External view

Remote in (9P)

Remote out (9P)

Figure 5.3 Sony nine-pin interface P2, used in many professional editing interfaces, picture and sound.

A standard analogue colour video signal has a bandwidth of 5.5 mHz and specialist techniques are needed to record this signal. An analogue tape recorder such as a domestic cassette recorder can only reproduce a range of 10 octaves whereas, to record video successfully, an 18-octave range is required. The only practical way of achieving this speed is by using a rotating head, which scans the tape at a high speed, while the tape itself runs at a slow speed to allow a reasonable amount of data to be stored on the tape. This is known as *helical scanning*. A control track is recorded to precisely define the position of the tracks written by the recording drum on the tape. This allows the head to be synchronized correctly to replay the recorded picture. Most digital audio tape recorders use the scanning head technology to allow their high-speed digital data streams to be successfully recorded.

The scanning head was first successfully used by Ampex in 1956, when it launched its 2-inch quadruplex – the first commercially successful video recorder. Video recording was developed further when 1-inch tape was scanned at a greater angle, allowing still images to be produced at tape standstill. Then, to protect the tape it was placed in a cassette. These form the basis of tape-driven video recorders today.

In the 1970s, videotape began to replace film in the broadcasting industry. A small format was developed that could match the quality and versatility of 16-mm film equipment in both size and quality. This was the helical scan U-matic ¾-inch tape cassette system, but with a linear speed of only 3.75 inches/second the audio quality was poor, as was picture quality. The U-matic format is now occasionally found as a viewing format. It formed the basis of our present VHS system.

In the mid-1980s, the Sony Betacam and the Matsushita/Panasonic MII camcorder video cassettes were introduced. Rather than recording combined (composite) signals, which included colour and brightness (or luminance), Betacam recorded the components separately in *component* form, a technique which resulted in better quality pictures. Betacam SP revolutionized location shooting in Europe and offered a viable alternative to high-quality 16-mm film production. The Beta SP format has been superseded by a variety of digital video formats from various manufacturers, each of which aims to serve a particular part of the market at a particular cost. Digital video ensures a higher quality recording, and on copying little generation loss. The *digital video* (DV) formats range from the 'amateur' Mini DV up to the Panasonic DVPRO50 format, which is used extensively in ENG work.

Video compression

The DV digital formats are all based on *video compression* techniques. Compression allows vast amounts of video data to be recorded in manageable ways.

A single frame of a broadcast TV picture needs about 265 million bits per second to define it. This volume of data can be difficult to manipulate in *data streams*, especially in operations like editing. The industry has therefore developed a number of *data reduction* techniques and, since digital data is just numbers, it lends itself to precise mathematical compression and decompression.

Analogue:		
Format	Use	Audio
Quad 2-inch	Archive broadcast	1–2 tracks analogue
U-matic (various)	Archive industrial/News	2 tracks
1-inch C	Archive broadcast	3–4 tracks
SVHS	Industrial	2 tracks analogue 2 tracks FM
Sony Betacam (SP)	Broadcast	2 tracks analogue 2 tracks FM
Panasonic M11	Archive broadcast	2 tracks analogue 2 tracks FM
Sony HDV1000	Archive HDTV broadcast	2 tracks analogue 4 tracks FM

Digital:		
Format	Use	Audio
JVC Digital S	Broadcast	4 tracks, 16-bit PCM
Sony DigiBeta	Broadcast	4 tracks, 20-bit PCM
Ampex DCT	Broadcast (archive)	4 tracks, 20-bit PCM
Sony D-1	Mastering (archive)	4 tracks, 20-bit PCM
Sony/Ampex D-2	Mastering (archive)	4 tracks, 20-bit PCM
Panasonic D-3	Broadcast (archive)	4 tracks, 16-bit PCM
Mini DV (aka DV)	Industrial/broadcast	2 tracks, 16-bit PCM
Sony DVCAM	Industrial/broadcast	2 tracks, 16-bit PCM 4 tracks, 16-bit PCM
Panasonic DVPRO 25	Industrial/broadcast	2 tracks, 16-bit PCM
Panasonic DVPRO 50	Broadcast	4 tracks, 16-bit PCM
Sony MPAG IMX	Broadcast	4 tracks, 24-bit PCM 8 tracks, 16-bit PCM

Figure 5.4 Professional videotape formats.

Sony		
XCAM (disc)	Broadcast	4 tracks, 24-bit PCM
		8 tracks, 16-bit PCM
Sony		
Beta SX	Broadcast	4 tracks, 16-bit PCM
(Note: low-budget theatrically released films may be made on the better quality of the above digital formats.)		
High definition digital:		
Panasonic		
DVPRO HD (D-12)	Broadcast/cinema	8 tracks, 16-bit PCM
Sony		
HDCAM	Broadcast/cinema	4 tracks, 16-bit PCM
Panasonic		
D-5 HD	Broadcast/cinema	4 tracks, 20-bit PCM
		1 analogue
Toshiba/BTS		
D-6	Post production	12 tracks, 20-bit PCM
Where no manufacturer is mentioned, more than two are involved.		
Note: Video standards are overseen worldwide by the ITI (International Telecommunications Union).		

Figure 5.4 continued.

The degree of digital data compression possible for video can be as low as 2:1 and as high as 100:1; 12:1 is the maximum compression used in broadcasting, 10:1 is used for Sony SX and a mild 2:1 for DigiBeta, but the resultant perceived picture quality after compression mainly depends on the amount of processing power of the computers involved, rather than the actual methods used.

The actual *digitizing* of a colour video signal requires the brightness (luminance) and the colour (chrominance) to be digitally sampled. The protocols are industry wide, laid down to an international standard (ITU 601). In DigiBeta, Betacam SX, Digital S and DVCPRO 50 the luminance is sampled four times to each chroma signal, which is sampled twice – the term 4:2:2 is used when referring to this digitization. High definition video systems such as the Viper FilmStream handle the uncompressed 4:4:4 quality required for cinema productions made on video. This system records directly onto hard disks – recording on to tape would mean continual tape changes, so great is the amount of storage needed. Computer editing systems can record to hard disks, producing on-line broadcast quality pictures either uncompressed, or with 2:1 compression and 4:2:2 processing. Disk-based cameras such as the Sony MPEG IMX disk systems use MPEG compression, with the advantage that recordings can be immediately edited and copied at speed when the disks arrive at the editing workstation. Some videotape formats are compatible with compression systems on disk, allowing for the transfer of data into a workstation from tape at a speed faster than real time.

For some applications, video needs to be transferred and edited at broadcast quality; however, for many types of production, less than broadcast quality will do – picture editing systems and settings are examined more fully in Chapter 9.

Film recording

Video can be successfully transferred or recorded onto film for cinema projection, but its quality depends on the quality of the original material. High definition formats produce excellent results and HiDef productions made for cinema release are becoming more popular.

Some recording systems use lasers to record on film, others use electron beams. Here the image is broken up into three separate colours, red, green and blue. The beam then records each line of picture three times, one for each colour. For a 2000-line resolution image (2K film), it records 2000 passes. The process is slow – it takes up to 8 seconds to record each individual frame. Laboratory film with high definition and slow exposure time is used (eight times slower than camera film). The results can be indistinguishable from camera-originated shots, as video-generated special effects demonstrate in blockbuster movies.

Audio on video recorders

Modern digital video recorders usually offer four or more channels of digital sound, with high definition formats offering eight. A high-quality digital audio channel requires a data rate much lower than that needed for video. Adding data to record an audio signal onto a video track increases the data rate, in total by only a few per cent. The audio can then be recorded by the video heads as small blocks of audio placed at specific intervals along the main picture tracks. At the same time, the audio sampling rate must be locked to the video rate. Since picture recording is intermittent, the recording does not take place in real time, but is compressed and expanded, or *companded*, to cover the time gap. As information comes out of the 'audio buffer' it immediately releases more space.

All digital formats have one thing in common: they provide digital audio that is of high quality and that can be copied many times without quality loss. Sixteen-bit quantization with 48 kHz sampling is the standard. However, some of the high definition formats offer 20-bit encoding.

The format chosen for a production will depend on the type of production. For high-end productions, DigiBeta or XDCAM might be chosen, or a high definition system. For news, DigiBeta is too bulky – DVCPRO or Beta SX might be more appropriate. For a 'fly on the wall' documentary, a small DV camera could be very suitable.

Viewing pictures in audio post production

Many computer-controlled audio workstations are able to record pictures as well as sound on their hard drives, with few interface problems. However, the image quality may well depend on whether the original picture was edited at full broadcast resolution or, as with productions shot on film, the picture edit is just a low-quality guide that will later be matched to the camera negative. Here, quality of picture is less important than the ability to maximize the amount of data held on the available storage devices.

A few audio workstations cannot currently import video – here a separate video machine has to be synchronized to the system. The format is not too significant: all that is important is that it can support timecode and be successfully interfaced – Beta SP decks are often used.

Video disc recorders can also be used to replay pictures – their quality depends on their cost. As with combined video and sound workstations, they provide instantaneous access to located points within the programme material, rather than waiting for tape to rewind.

Viewing images

Pictures viewed in the audio post production suite can be shown on a monitor, projected via a video projector, or via a film projector. Images must be easy to see, with good definition and, if projected, well lit. Video pictures can be successfully projected onto a screen, but quality of image is very cost dependent. Often, in television audio post production picture quality depends on the quality of the video material available, which may just not stand up to gross enlargement. Although impressive, projectors do not recreate the standard domestic TV environment, and the poor definition and reduced environmental light levels necessary to provide a bright picture can lead to eye strain. Television monitors are often used in studios where television projects are being mixed. Monitors are graded according to quality, with grade 1 being used for checking broadcast transmission quality.

In a large film re-recording theatre, mixing suite or dubbing theatre, a film projector may be needed to produce sufficient light output for the screen. They can be especially adapted to run both backwards and forwards, but are more likely to be used only for checking the final film print against the final mix, rather than throughout the premixes and mix. To assist the re-recording mixer, a timing counter is displayed, under or within the screen, or on the computer workstation screen. This must be clear and easy to read, measuring time in timecode or film feet and frames. For video operations, the timecode may be superimposed onto the picture using the switch on the playback VTR.

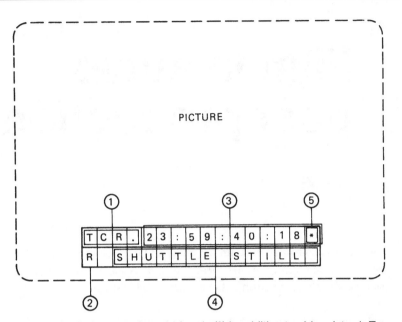

Figure 5.5 A monitor displaying superimposed code (3) in addition to video data. 1, Type of time data; 2, videotape recorder assignment; 4, videotape recorder status; 5, indication of vertical interval timecode.

Comparing film and video

Video and film are very different technologies. Film is mechanically based and comparatively simple to understand, while video is dependent on complex electronics, but both systems have their own attractions.

Film has been established for over 100 years and is made to international standards accepted throughout the world. Indeed, it was once the only medium of programme exchange in television. This changed with the introduction of the videotape recorder, and in particular with the introduction of the broadcast video cassette format. Film stock, processing and printing are expensive processes, and film is now a specialist quality format used in cinemas and for high-budget television production. For location shooting, 35 mm is difficult to match in terms of contrast ratio and definition, and film will still be used in theatrical productions for some years to come. But the cost of stock and equipment will increase as fewer productions are made on film.

The projection of film images remains an excellent way of displaying images, but high definition video can produce comparable results, which may well be transmitted directly to a motion picture theatre in the future. The saving in film stock costs and the cost of transporting film cans to theatres will encourage the digital distribution of productions made for theatrical release.

In the next chapter we look at specialist film technology in more detail.

6 Film in audio post production

Tim Amyes

Film is now only used as an acquisition medium for studio and location shooting, and as a projection medium for theatrically released films. Between these two points – acquisition and transmission – film travels a digital path, passing through a string of computer-based systems.

Film is used on high-budget productions, where the considerable cost of the film stock and the associated processing is acceptable. Film can give higher contrast and better definition than comparable high definition video systems. But video systems are continually improving and, to counter this, so is film technology – offering film stock of high speed with very little grain.

However, 16-mm film, which is used in European television for high-end productions, will no doubt be supplanted by high definition video. For cinema releases, 35-mm film is still the preferred shooting format with a release print 'struck' from the film camera negative. However, some features follow the 'digital film laboratory' route. Here, the original camera negative is transferred at highest quality (2K) to video and the film is 'discarded'. The film release print is then produced directly from the edited video master using a *film recorder*. This method allows access to a wide range of digital special effects during editing, digital colour grading and perhaps a speedier turnaround in some areas of production.

In film acquisition, audio has to be recorded on a separate sound recorder, a DAT machine or a hard disk recorder. On location, each shot is marked using a slate or clapperboard, with timecode. A long 'pre-roll' time may be needed for later synchronization. After the camera negative has been developed, the film is transferred to video so that it can be edited in a non-linear editing system.

The film *rushes* or *dailies* are transferred to video in a telecine suite. Sometimes all the rushes are transferred, sometimes only *selected takes*. Sometimes the sound is transferred and synchronized with picture at the film transfer stage; alternatively, this job may be done later in the cutting room. In this scenario, the rushes will be synced up by the assistant editor in the NLE.

For cinema release the operator will work with a print or copy of the negative and make a video copy as a 'work print' for picture editing. For television transmission he or she may work with the original negative, producing a high-quality digital video copy that, once edited, becomes the transmission master.

Synchronizing the sound and picture rushes in telecine has to start by noting the first stationary timecode slate visible on the film at the point where the clapperboard just closes. The telecine machine is held stationary at this point and its controller records the picture frame position. The operator then types this timecode number into the synchronizer, which controls both audio and telecine machines. He or she finds the appropriate audio tape or sound file and, using the timecode first identified on the picture, locates to the slate number on the audio. Both machines now have the necessary synchronizing information. The picture slate has just closed, the audio sits on the stationary timecode number identified as the clapper 'bang', and on command both machines synchronize.

If tape is used, then both machines will reverse and then go forward and lock up. The telecine machine will use its sprockets to identify the correct synchronizing point and speed, and the tape machine will use timecode. For this to work there must be 10 seconds or so of timecode recorded on the audio tape before the slate closes. This will allow the synchronizer to run the tape and film back and then forwards to run up to synchronization. If there is a break in code, the synchronizer will be confused and may 'run away'. The sprockets on the film are continuous, so even at shot changes they still provide synchronizing information. The system can be completely automated with film recording timecode systems.

In some productions the camera original negative film may well be transferred without sound – *mute*. Perhaps, again, only certain takes that are needed will be transferred – *circled takes*. At the editing stage, the sound rushes are synchronized in a similar way to the telecine process, but using the autosync function offered in the workstation.

Although telecine machines are 'kind' to film, it is unwise to run camera negative through the machines more than is necessary. There is always a danger of scratching an irreplaceable film camera original, so copies or prints are used if possible. Synchronizing sound with original negative here is not advised – running film backwards and forwards through a film gate to find synchronization may well damage the original negative film.

To speed up synchronizing the audio rushes with tape, a sound recordist may record only synchronous camera material to the main record machine. A second recorder is used for wild tracks and guide tracks. Thus, each picture roll will have its own corresponding sound roll. If only one tape is used on location, once the synchronous transfers are completed, the additional sounds will be added to the end of the picture roll. Good logging is essential on location, for initially the picture editor will 'discard' effects and sounds that are not needed in the picture edit. The sound editor will need to find these later.

Film release

Once the film has been transferred to video it may never need to be returned to celluloid and no relationship between the camera negative and the video need exist – this is the path of productions originated on film, but destined purely for television.

Alternatively, if the film is to be shown in the cinema (and the digital film lab route is not taken), it is essential that each frame of camera negative can be identified against each edited video in the editing system – so that an accurate print can be produced. Careful planning in the early stages of transferring the negative to video avoids any costly mistakes that may only be discovered once the final print has been made. At this point, either the soundtrack will have to be altered ('pulled into sync') or the negative cut will have to be altered; this will entail expensive reprinting. There are various ways in which timecode and film stock manufacturers can help to ensure accuracy of the negative cut.

Although camera film can be timecoded, a more widespread system is to assign each frame a number. When a photographic negative is produced, the numbers are exposed along the edge of the film (every 1 ft for 35 mm, every 6 in for 16 mm), which becomes visible when the film is developed. These numbers are known as *key code* or edge numbers. Each number uniquely identifies a frame of camera original. Edge numbers can be read by eye or by a machine using a bar code reader. They identify frames in relation to the position of the code (e.g. code XXXXX plus eight frames). These machine-readable key codes are invaluable in film projects that will eventually be printed up for theatrical release.

Conforming film

Film is not physically handled in post production; instead, video copies are used. Edit decision lists and edge or key numbers are used to conform camera negatives to the edited video picture. It is

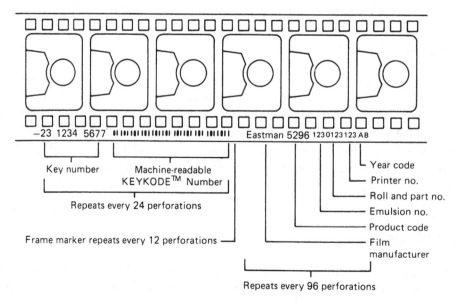

Figure 6.1 An example of Edgeprint format (courtesy of Kodak Ltd).

essential that, before any editing begins, each frame of film is identified and related to its appropriate frame of video. The simplest way of doing this is to give each roll of film a timecode sequence directly related to the edge numbers. Since the film is physically sprocketed, a time start point can be identified at a specific frame: this also becomes the timecode start on transfer to video. Each roll of camera original must have unique numbers.

At the start point (for example, 01.00.00.00 time) the key number is read off – as the sprockets pass it is possible to read time and mathematically relate it to the edge numbers: all that is necessary is to know the start point time. This is usually identified by punching a hole in the film. To conform the film, the timecode readings are converted to edge numbers, producing a *film cut list*. The negative can then be cut frame to frame against timecodes. Alternatively, as the original video transfer takes place, the bar code edge numbers are machine read automatically by the telecine machine and logged and correlated against the video's timecode. This uniquely labels every frame of film. The sync sound for the project is conformed later using the video transfer as the master. The original timecodes are maintained throughout the picture editing. On completion of the edit, these timecodes are tied back to the keycodes, allowing the negative to be matched to the video edit frame for frame. In NTSC systems there are additional problems because the *pull-down* mode needs to be taken into account. There can be similar problems in countries using PAL when 24 fps 35-mm film for cinema release is transferred to PAL workstations. The practical implications of working with differing frame rates and the concept of pull-down are discussed more fully in Chapter 9.

Film timecode

Timecode can be successfully recorded onto photographic camera negative and various systems have been developed. These provide specific frame identifications for conforming final edited video pictures to original camera negative.

Arriflex (Arri) code system

In this system the SMPTE/EBU timecode appears in digital form along the edge of the film. The system uses one light emitting diode (LED) either in the gate or within the magazine – where it has to compensate for the film loop.

In operation the camera's timecode generator is synced to an external generator, which also feeds the audio recorder. Arriflex timecode and audio timecode are the same, and the transfer to video of both sound and camera negative is automatic. Again, one sound roll should correspond to one camera roll for speedy operation. As synchronization takes place, the Arriflex timecode derived from the negative, and the manufacturer's own bar code identifying each frame of picture, is held in a computer database or *flexfile*, together with the videotape timecode. When the video edit is completed, the timecode on the video is referred back to the film manufacturer's bar code, producing a film cut list. The negative is then matched up and printed to produce the final print, with its timecode providing confirmation of print to video synchronization.

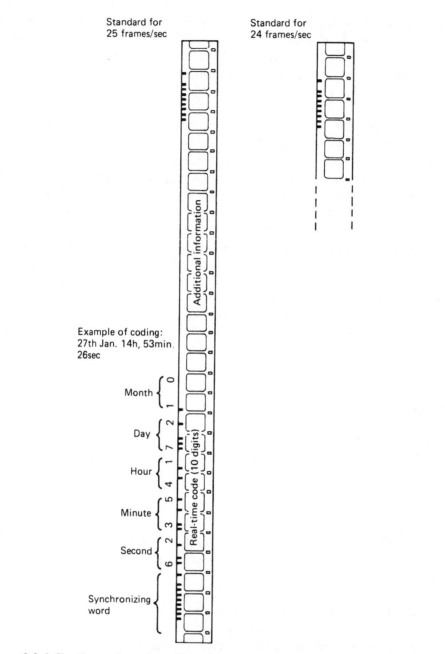

Figure 6.2 A film timecode system. There is no international timecode system in film (courtesy of F. Berstein).

Aaton system

The Aaton system was specifically designed for 16-mm use. Both digital and visual readable numbers are printed onto the film stock in the camera using a row of seven red LEDs. The system has been developed further for use with 35-mm film. Information recorded includes timecode with visual information covering frame, scene, take and roll, and production number with the date of the shooting. The negative is now prepared – the shots are identified from the list and then cut together with any special visual effects required. The completed negative is then ready for printing with the soundtrack.

Sound on sprocketed film

Sound can be recorded photographically or magnetically on sprocketed film, usually referred to as *mag stock*. Sprocketed magnetic film is an oxide-coated film of the same dimensions as camera film. Sound is recorded in the analogue domain. Like telecine machines, spocketed magnetic machines (dubbers) are driven in synchronization by a low-voltage biphase generator. Until the 1990s, all film soundtracks were mixed onto magnetic film. Sixteen-millimetre sprocketed magnetic film offers one or two soundtracks (16-mm sepmag). In 35 mm, the soundtrack is positioned on the left-hand side of the film, in the same position as the soundtrack on a final theatrical release print. Up to three tracks can be recorded across the film in normal usage, although it is possible to record four or even six tracks. Mag is now only used as an archive format.

Photographic film recording

Sound recording onto sprocketed film is now restricted to the highly specialized photographic methods used for cinema release prints. Here, the sound is replayed from an optical soundtrack produced by an optical sound camera. As cinema sound reproduction systems are of proprietary design, it is necessary to obtain a licence for their use. This applies to all digital sound systems.

Cinema exhibitors are wary of the cost of new digital surround sound systems, which may or may not bring in increased revenue. Although over 60 000 screens have converted to digital sound worldwide, most have adapted their installations to use a maximum of six sound channels. The analogue optical

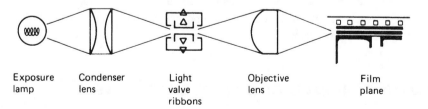

Exposure lamp   Condenser lens   Light valve ribbons   Objective lens   Film plane

Figure 6.3 An optical recording system producing a variable area soundtrack. The light valve ribbons open and close within the magnetic field.

stereo soundtrack still remains a universal standard throughout the world. It can be played off any 35-mm projector. Even the oldest sound head of a mono machine will successively reproduce audio! It can be used as a backup if the digital system fails.

The analogue photographic recording system is well suited to stereo, since two separate tracks can be easily recorded onto a photographic emulsion. These tracks are either Dolby A or SR encoded, the format being called Dolby Stereo Bilateral Variable Area, or just Dolby SVA. The use of a sound matrix within the Dolby recording system allows four separate sources to be encoded onto the two tracks. Three are for use behind the screen (left, centre and right) and the fourth (surround channel) is located in the audience area.

To record sound photographically, light-sensitive, high-contrast film is drawn across a slit formed by two metal ribbons held in a strong magnetic field. This 'light valve' opens and closes as the signals are fed into it. The light shines through this slit from an intense quartz halogen lamp and traces a path onto the photographic film; the light varies in intensity with the recorded signals. This standard soundtrack is located near the edge of the film, just inside the sprocket holes.

When the film is processed and printed, the photographic image of the soundtrack is exposed as a white-on-black line of varying thickness. Since the areas of the black and the white within the soundtrack are continually varying, this is called a *variable area soundtrack*. The soundtrack is replayed by being projected, not onto a screen but onto a photoelectric cell, which picks up the variations in light and converts them into electrical signals. This reader must be well set up, otherwise the bandwidth of the system will suffer.

The analogue photographic soundtrack has limitations. The soundtrack has a frequency response that is substantially flat up to 12 kHz (only an adequate figure), equal to that of 14-bit digital sound with a sampling frequency of 32 kHz. However, it is substantially free from compression effects at high level, and at modulations of up to 100 per cent performs in a perfectly linear manner; beyond this (like digital audio recording) it can go into heavy distortion. To extend the frequency range, the track may be recorded at half speed.

Analogue recorded optical tracks are of reasonable quality; much of the poor sound associated with them is due to the way in which they are handled – rather than any major defect in the recording system. Audio problems usually relate to high noise levels, which will increase with use. A film soundtrack can become easily damaged as it passes through a projector; the clear areas of the film become scratched, which causes clicks and background hiss on the loudspeakers. Noise reductions techniques can reduce these problems quite dramatically.

Recording analogue optical soundtracks

To assist the recording engineer in achieving a maximum level without distortion on the optical soundtracks, two indicator lights are often provided on the recorder in addition to meters: one flashes yellow on a clash of the light valve, indicating 10 milliseconds of distortion, which is considered inaudible; one flashes red on clashes over 100 milliseconds, indicating audible distortion.

To ensure that the sound is exposed successfully, *sensitometry tests* are made through the recording chain so as to determine:

- the optimum exposure or the amount of light used in the optical camera;
- the processing time needed to produce the right contrast, creating good definition between the dark and clear parts of the film, both for negative and print material.

Due to the nature of the photographic process, it is impossible to obtain the ideal 100 per cent white and 100 per cent black intensity on a film soundtrack, so a compromise has to be reached that will produce good sound quality and low background noise. When the sound is printed onto the release print, the image should have exactly the same definition as the original camera recording. However, the system is not perfect and there is likely to be some image spread, which will result in distortion and siblance.

Digital optical soundtracks

Dolby's digital optical sound system allows six discrete digital tracks to be recorded onto the film: screen left, centre and right, with sub-bass behind and left and right surround in the auditorium. The

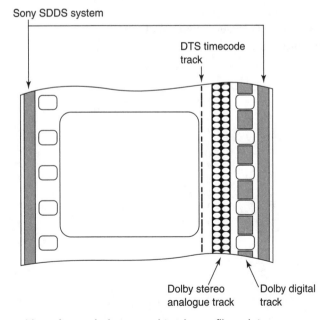

Figure 6.4 Relative position of encoded surround tracks on film print.

system utilizes black and white clear pixels, which are optically recorded by laser and which sit between the perforations on the 35-mm film and use Dolby's AC3 coding system.

The same speaker layout is adopted by Digital Theatre Systems (DTS), a system which first appeared in 1994 using CD-ROM technology. An optical track on the film carries timecode data that also contains reel numbers to allow the CD-ROM to follow the film (CD-ROMs cannot be played on an ordinary CD player). This code track is photographically printed between the picture and optical tracks. It offers added flexibility since foreign versions may not require new projection prints, merely replacement discs. In the Sony Digital System (SDDS), blocks of clear and cyan pixels are recorded continuously down both of the outside edges of the film. It uses the same ATRAC technology found in Sony's MiniDisc recorders. The audio information from the track is decoded to give left, left inner, centre, right inner, right and sub-bass behind the screen, with again left and right surround in the auditorium.

All these photographic systems, both analogue and digital, can be recorded onto one piece of film, although many passes in the optical camera are required. Very few facilities houses and film laboratories in the world offer photographic optical recording.

Optical sound is:

● easily copied by the photographic process;
● inherently distortion free (but goes into very heavy distortion on over-modulated peaks);
● susceptible to poor handling;
● of adequate quality unless digital;
● produced to universal standards;
● entirely mono compatible in analogue stereo SVA format;
● capable of being produced in multiple formats.

Part 2

The Post Production Process

<table>
<tr><td>7</td></tr>
</table>

7 Post production workflows

Hilary Wyatt

In Chapter 1 the concept of a linear post production workflow was introduced, and indeed most productions follow this path. Partly, this is a result of the way in which production technology has evolved, and partly it is a result of the need to use production resources efficiently to stay within budget. On most productions, any overlaps in the various stages of post production are minimized as far as possible, so that, for example, the sound editors will only start work once the picture edit has been finalized. An exception to this may be made on medium- to high-budget feature films, where the picture editors may work concurrently with the sound team so that temp mixes can be produced before the picture is locked.

Although working practices vary from production to production, this short chapter looks at a few typical workflows, prior to examining the process in some detail in the remaining chapters.

In the workflow shown in Figure 7.1, the picture edit, tracklay and mix are carried out by the picture editor using the facilities within the picture editing system to output a broadcast master. In this scenario, the pictures are edited at broadcast or *on-line* resolution, and a few sound effects and music tracks are added to the production sound, which will be mixed with little or no eq, filtering or

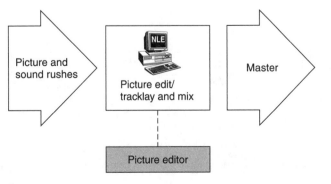

Figure 7.1 Workflow 1.

processing. This method of production is suitable for projects that have relatively simple soundtracks, such as low-budget documentary and reality TV, as well as fast turnaround projects such as news, sports inserts and promos.

In the workflow shown in Figure 7.2, the picture editor will edit the picture, do some preliminary work on the production sound, and perhaps lay some sound effects and music in the edit suite. These tracks will then be handed onto a sound mixer, who may augment the sound effects and music tracks with material from the studio library, record voice-over to picture, and use eq, filtering and effects processing to improve the quality of the production sound. This method of production is suitable for projects that require a 'fuller' mix, and where the skills of the sound mixer are needed to get the most out of the production sound, balancing it against other elements in the mix to produce a dynamic soundtrack that conforms to the appropriate broadcast specifications. Most studio-based productions (such as quiz shows, game shows, soaps), low-budget TV drama, commercials and some documentary work will be produced in this way. Some productions intended for overseas sales will require an M&E (Music and Effects) track to be made once the mix is complete. In simple terms, this is the mix minus *any*

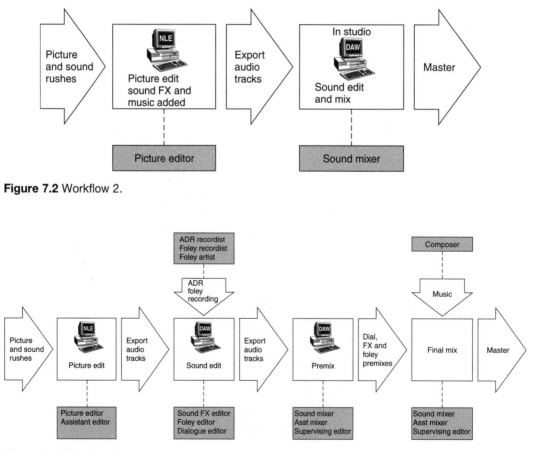

Figure 7.2 Workflow 2.

Figure 7.3 Workflow 3.

dialogue – this enables the production to be dubbed into a foreign language at a later stage. An exception to this is documentary, where production sound dialogue is usually left in the M&E, although the voice-over is taken out. Here the voice-over will be dubbed into a foreign language and the production sound dialogue will be subtitled. Depending on the running time of the production, the amount of time allowed for the mix may be anything from a few hours to a day per programme episode.

In the workflow shown in Figure 7.3, the picture editor hands the project over to a sound editor, or team of editors, who will prepare or *tracklay* the dialogues and sound effects to a high standard in readiness for the mix, and oversee the recording of foley effects and ADR. Because of the higher number of tracks that result from this process, the sound mixer will often premix the dialogues, sound effects and sometimes the foley to reduce the number of decisions that will need to be made in the final mix. At this point, the music, which would probably be specially composed or *scored*, is laid up using the timecode cues provided by the composer. The sound mixer then produces a full mix consisting of the dialogue, sound effects, music and foley. This method of working is used on higher budget TV projects, such as drama, high-end documentary, animation and low-budget feature films. Because of the additional work involved in the tracklay and mix, these are usually scheduled over a number of weeks rather than hours or days. There is usually a need to produce an M&E as well as other versions of the mix for international sales and archiving purposes. These versions are known as the *delivery requirements*, and are produced once the final mix has been completed.

The workflow shown in Figure 7.4 involves a much larger crew, who will work over a period of weeks (often months) to produce a highly detailed soundtrack using all available resources. As a result, this type of approach is only used on medium- to high-budget feature films. Often, some members of the sound editing team will begin laying *temp tracks* whilst the picture is still in the process of being

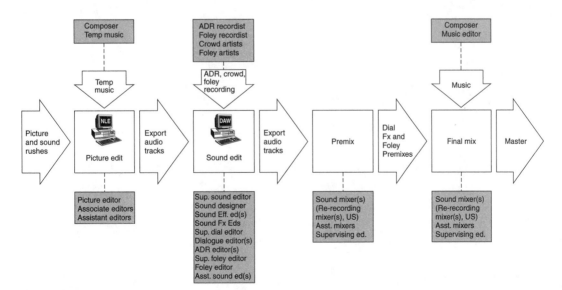

Figure 7.4 Workflow 4.

edited. These temp tracks will be used to create a *temp mix*, which may be used in audience research screenings arranged by the production. This approach is perhaps the least 'linear' of all, because picture editing, sound editing and mixing may take place concurrently until the final version of the film has been approved.

The 'hierarchy' of editors will also include a music editor, who may lay temp music tracks early on in the schedule and assist the composer in delivering the completed score to the final mix. The mix itself will involve a very high number of tracks, which will be premixed and then final mixed by one or more sound mixers or *re-recording mixers* (USA). Because most feature films aspire to international distribution, the delivery requirements can be quite complex, and these will be produced once the final mix is complete.

The remaining chapters in Part 2 examine the processes involved in post production in more detail.

8 Recording audio for post production

Hilary Wyatt

Aims

Audio that will be handed over to post production needs to be recorded in a way which enables it to be edited and mixed successfully. Ideally, some advance communication between the recordist and the post production team is desirable, but not always possible due to schedule constraints. Failing this, the recordist needs to think ahead and ensure that the post production department receives the material it needs to do its job well. This should include:

● Reproducing the original location performance as faithfully as possible.
● Minimizing audio quality changes within a scene, from shot to shot.
● Reducing extraneous noise on set/location.
● Recording dialogue and fx wildtracks.
● Recording with as little reverb as possible – this can be added later but not removed!
● Avoiding dialogue overlaps, which may create difficulties in editing.
● Recording on the appropriate medium, format (mono/stereo) and sample rate/bit depth.

In practice, it is important to understand that a recordist may be compromised in achieving a good-quality recording by very tight schedules and difficult recording conditions. However, it is useful to understand the significant factors that will affect the overall quality of the sound.

Types of microphone

The most common types of microphone used to record audio for film and television are outlined below.

Boom mics

There are essentially three types of boom mic:

1. The *omnidirectional* mic is the most reverberant field sensitive. It picks up sound equally around the 360° axis, and is used in situations where directionality would be a disadvantage – for example, in recording a group of people seated around a table. It is also the least sensitive to wind and handling noise, and so is a good choice when a mic needs to be used hand held.
2. The *cardioid* mic has a kidney-shaped response pattern and is most sensitive to sounds in front of the mic. Sounds around the sides of the mic are picked up at a much lower level, whilst sounds from the rear barely register. This means that the cardioid offers a degree of directionality, as well as the ability to pick up some degree of ambient noise. A *supercardioid* mic is less reverberant field sensitive and has an exaggerated response towards the front of the mic. Cardioids should be used in situations where it is important to record the natural ambience of a location. They can also be used as fixed mics because their response pattern makes them relatively forgiving if the artist moves slightly *off axis*.
3. The *rifle* or *shotgun* mic is often used in film/TV location work. Because of its highly directional *hypercardioid* response pattern, it is usually mounted on a fishpole (see below) operated by a *boom swinger*, whose job it is to keep the mic aimed precisely at the subject, who must remain *on axis*. Rifle mics tend to record a drier sound, but are useful in noisy locations as extraneous sound is less likely to be picked up.

Bidirectional mics have a figure-of-eight response pattern, picking up sound equally well from behind the mic as well as in front. This response pattern makes them unsuitable for use as a boom; however, bidirectional mics are used when recording in M/S stereo (see below).

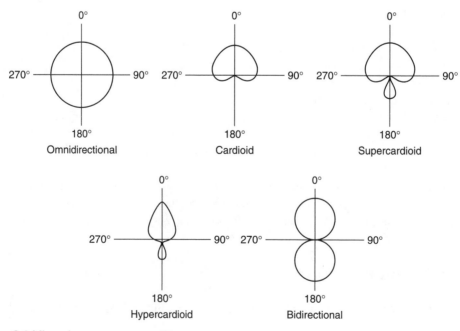

Figure 8.1 Microphone response patterns.

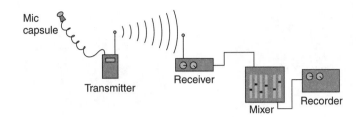

Figure 8.2 A radio mic set-up.

Personal mics

Also known as *tie, lavalier* or *lapel mics*, these are small microphones that can be attached to the artist in some way, or concealed in a nearby prop. Personal mics are always omnidirectional, although they do not reproduce natural acoustics very well and have limited dynamic range. They are also particularly prone to pops and wind noise. The main advantage of a personal mic is that it remains in close proximity to the artist regardless of movement. However, this can also be a disadvantage: whatever the length of shot, the sound always comes from a close perspective, which may seem unnatural on a medium to wide shot. As personal mics are *condenser* mics they require a pre-amp, which must be powered either by mains or battery.

Personal *line mics* can be directly wired to the production mixer or camera, in which case the movement of the artist is restricted only by the mic cable. For some productions, however, complete freedom of movement is required and in this case the answer is to use a *radio mic*. Here, the mic is powered by a battery pack carried on the artist's person, and the signal is sent to the production mixer or camera via an RF transmitter. Radio mics are very useful for capturing sound under difficult recording conditions involving physical activity and spontaneous action (some mics and battery packs are waterproof, for example). In film and drama work, they are often used to cover wide-angle shots, where it is not possible to get the boom mic close enough to the subject without it being seen by the camera. Radio mics can be prone to electrical interference and are, of course, reliant on a ready supply of spare batteries.

Mono and stereo recording

Mono

Where a scene is predominantly dialogue driven, the recordist should obtain the cleanest recording possible for post production. To this end, mono recording using a directional mic is standard practice. With the exception of crowd recordings, dialogue for film and TV will be mixed in mono. Occasionally dialogue is recorded as M/S stereo, but the stereo (side) component of the signal is generally discarded in post, as it usually consists mainly of ambient and often unwanted noise. This leaves the mono (mid) leg as the track used in the mix. A further difficulty of stereo dialogue recording is that the stereo image will move around from shot to shot each time a new camera and mic position is set up. In the case of

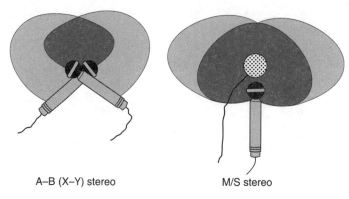

A–B (X–Y) stereo   M/S stereo

Figure 8.3 Stereo recording techniques.

a scene composed of reverse angles, the stereo image will shift from one side of the screen to the other, once both angles are cut together. Some spot fx that are recorded as wildtracks (such as doors, phone rings, etc.) can be recorded in mono, as they will be tracklaid and mixed in mono (stereo reverb may be added later).

Stereo

In order to produce a stereo recording it is necessary to use a stereo mic, which is actually comprised of two mics housed in the same capsule. There are two types of stereo recording:

1. *M/S* (Mid and Side) recording uses a *coincident pair* of mics. A directional mic (usually cardioid) picks up the mono (or mid) part of the signal from straight ahead, whilst a bidirectional mic with a figure of eight response pattern picks up sound from the left and right (or side). The advantage of this method is that the width of the stereo image can be controlled in post by adjusting the amount of side. However, the signal must be passed through a matrix in order to decode a left–right stereo image. For this reason, it tends to be unpopular with sound editors and mixers, and is rarely encountered in post production.
2. *A–B* (or *X–Y*) stereo uses a coincident pair of cardioid mics placed at 90° to each other to produce a left–right image, and positioned at 45° to the subject. This method is often used for recording sound fx and atmos, and is preferred by sound editors and dubbing mixers. In general, stereo recording should be limited to atmospheres (such as sea wash, big crowds, traffic) and action (such as moving vehicles).

Microphone position

On-camera microphones

Built-in camera mics on camcorders are very often omnidirectional, which means that sound is recorded in an unfocused way, together with any extraneous noise. Professional cameras can be fitted with a choice of mic (usually a shotgun), which means that whatever the camera is looking at will be *on mic*. In either case, the mic is, of course, tied to the position of the camera and all audio will be recorded from the same perspective as the picture. This may mean that audio quality will vary from shot to shot, posing a continuity problem when the material is edited together, and in wider shots it may be difficult to isolate foreground speech from the background noise. One solution is to record close dialogue with a personal mic and mix this with the output of the camera mic, which will record the natural acoustic of a location more successfully.

The advantage of using an on-camera mic is that, in some situations, such as in covert or observational filming, it seems less intrusive than fitting the subject with a personal mic or following them round with a boom. On productions where crew levels are an issue, this option means that the camera operator can record both sound and picture. In fast turnaround situations such as news gathering, the audio can be routed directly into the camera inputs. However, the disadvantage of this is that the camera's *Automatic Gain Control* must be used to control recording levels, which may result in an audible 'pumping' effect. A better alternative is to route the audio to the camera inputs through a small dual input mixer, which can be attached to a belt for ease of use. Once the shoot is complete, the location tapes can be sent directly to the edit suite.

Off-camera microphones

For most professional TV and film productions, microphones will be sited off camera in the optimum position for sound quality, regardless of the camera's position. In order to 'cover' a scene or location, a number of mic types may be used in a single set-up. However, the position of each microphone in relation to the subject is also a critical factor in providing the editing department with usable sound. The main techniques used in film and TV can be summarized as follows.

Mic with fishpole

This is one of the most common techniques used in location work involving a sound crew, mainly because it achieves good-quality sound and allows the subject freedom of movement without being hampered by mic cables or fixed mic placement. The mic is mounted on a long *boom* or *fishpole*, which enables the mic to be held just out of shot. This technique requires a dedicated crew member known as a *boom swinger* to follow the subject around whilst keeping them on mic. To record speech, the mic is best angled down towards the subject from overhead, a technique which results in the most natural sounding dialogue. Some ambient fx will also be picked up, resulting in a recording that reflects the acoustic properties of the location in proportion to the dialogue. Audio 'perspective' can be suggested by lifting the mic higher for a wide shot, resulting in a thinner sound. A more intimate sound can be achieved by holding the mic closer to the speaker for a close-up. In some shots it may be necessary to mic from below, although in doing so, it may be harder to keep the mic out of shot. This technique tends to make for a boomier recording which reflects the mic's proximity to the speaker's chest. It may also mean that fx such as footsteps are more prominent in the recording than might be desirable. To reduce wind and handling noise, the mic should be suspended in a *shockmount* and housed in a *windshield*.

Fixed booms

In situations where the subject is not likely to move around much, a fixed mic stand can be used. This technique is limited to productions such as music and entertainment shows, simply because in order to be close enough to the subject, both the stand and mic will be in shot.

Fisher booms

The Fisher boom is a large platform boom designed for use within the studio. It can be suspended over a set, out of shot, and needs two operators: one to physically move the platform, and one to control the angle and placement of the mic over the set during takes. The advantage of these booms is that they can reach right into the back of a set without getting into shot, but are only suitable for large studios that have the floor space to accommodate the platform.

Spot/plant mics

Boom and personal mics may be hidden within the set/location itself. For example, in a restaurant scene the mic may be hidden within a table lamp or a vase of flowers. In choosing where to plant a mic, it is important to work out exactly where the performers will deliver their lines, and ensure the mic is directed at them. It is also important not to hide the mic in a prop that a performer is likely to pick up or move!

Slung mics

Mics can be suspended over the subject, a technique that is particularly useful for recording static groups of people such as studio audiences, orchestras and choirs. Its disadvantage is that the mic(s) cannot be adjusted during a take.

Hand-held mics

Hand-held mics are limited to news and other productions where it is acceptable (and sometimes even part of the convention) to have a mic in shot. An omnidirectional mic is often used and this can be moved around by a presenter without the need to remain on axis in order to obtain a good recording. How good the recording actually is will depend on the presenter's mic skills rather than the sound crew.

Personal mic worn on body

In some productions, personal mics are worn visibly in shot, on the lapel or pinned to the chest. Where line mics are used, the cable is hidden within clothing, with the mic connector led down to the ankle.

Most problems start when the mic has to be hidden on the performer's person, perhaps in the costume or even in their hair. The costume can be made to conceal the mic without actually abrading the mic in any way. Because they are so susceptible to wind noise, these mics are often buried deep within the costume, at the expense of HF response. Dialogue recorded in this way can sound dull and muffled, and will often need to be *brightened* in post production. Clothing rustle and wind noise are the main reasons why dialogue recorded with personal mics is often ADR'd, and the recordist may resort to tricks such as using loops of blue tack over the mic capsule to avoid contact with the costume.

Radio mics worn on body

Again, satisfactory concealment of the mic capsule without interference from clothing is the main issue. However, the battery pack and the transmitter antenna must also be concealed in a way that is considerate both to performance and movement. Unfortunately, once a take has started, the sound recordist has little control over the quality of audio recorded.

Using multiple microphones

Apart from the simplest set-ups, such as a news *piece to camera*, most film and TV productions will use of a number of mics to cover a single set-up. This is done for two reasons:

1. A scene or location may involve a number of actors, all of whom need to be on mic. In film or TV drama rehearsals, a set of *marks* (marked out by tape on the floor) should be established for a scene so that the actor(s) can replicate the same movements around the set on every take. These marks can be used to plan out the best mic positions, so that the actor effectively walks from mic to mic. In this scenario it is important that directional mics are used, and spaced correctly to avoid a sound being picked up by more than one mic. Where this does happen, the sound reaches each mic at a slightly different time. When the two mic outputs are mixed together (see below), they will be *out of phase* and *phase cancellation* occurs, causing the recording to sound oddly reverberant, and rendering the sound unusable in its mixed form. To minimize the chances of this happening, mics should be placed as close as possible to the subject. A useful rule of thumb is to measure the distance between the subject and the first mic. The next nearest mic should be placed at least three times this distance away from the first. Where a number of radio mics are used, each should be set to operate on a different frequency.

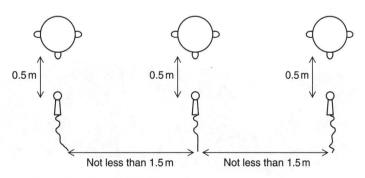

Figure 8.4 Using multiple microphones: the 3–1 rule.

2. The sound recordist may choose to record a scene using more than one different type of mic. This gives the sound editors some choice as to which they will use in the mix. For example, radio mics may be used on the establishing wide shot of a scene, where it is difficult to conceal a boom. On the subsequent medium and close-up angles, the recordist may switch to a boom mic/fishpole to obtain a better quality recording. This, however, will pose a problem in post production as the quality of the two mics will differ hugely and will not easily blend together in the mix. The wide shot may be marked down for ADR anyway, using a boom mic to match the rest of the scene, but the sound recordist may give the editors the whole scene recorded on both radio mics and boom for continuity. It is important that the recordist uses matching mics as far as possible, so that the various mic sources can be blended together in the final mix to sound as if they come from a single source.

Production mixing

In order to control the level of recording during a take, most sound recordists or *production mixers* will route all mic outputs to the record machine through a mixing console. The mixer can be a simple dual input mixer (see Figure 8.5), controlled by an ENG camera operator.

For drama and film location work, the recordist is likely to have a mains or battery powered portable mixer mounted on a custom-built trolley (see Figure 8.6).

Studio-based recordings will be routed through a full-sized console with multiple mic and line inputs. The sound recordist uses the console to do a number of tasks, which will improve the quality of the sound rushes delivered to post production.

● A number of mic sources can be balanced and mixed down to the number of tracks available on the recording format.
● A mix plus a separate recording of each mic input can be recorded assuming sufficient tracks are available on the record device.
● The sound recordist can achieve the optimum level on a recording by *riding the gain* using the console faders.
● Recordings can be equalized or compressed if necessary.

Figure 8.5 Dual input mixer (courtesy of SQN Electronics Ltd).

Figure 8.6 A production mixer on location (courtesy of Brian Simmonds).

- The faders can be opened and closed during a take, so that each mic channel is only open when required. This is preferable to leaving all mics permanently open, as it reduces the chance of phasing or picking up extraneous noise.
- A slate mic can be assigned to one input so that the take ident and clap are sufficiently audible at the head of each take.
- Headphone feeds can be sent to the director, boom swinger, etc. The recordist should also monitor the mix on headphones and check for noise or distortion.
- Line-up tone can be generated and recorded at the head of each take.

Studio and field recorders

The output of the mixing console is routed to the input channels of the record machine. Where a mixer is equipped with multiple outputs, the mix can be fed to one or more recording devices. Where a multitrack recording format is used, the sound recordist will record the mix on one track, and in addition may record each mic output separately onto spare tracks. In this case, the picture editor will use the single track mix to edit with: most editors would find it difficult to work with what may be between four and 10 tracks of audio at any one time. The separate mic tracks can be accessed later by the sound editors, who may need to use them in circumstances where the mix track cannot be edited together successfully (e.g. where dialogue overlaps or phasing occurs).

Recording to a single camera or VTR

The simplest way to record sound is on the same medium as the pictures, which eliminates any concern about sync between the two later on in post production. Most formats used on the shoot, or *acquisition formats* as they are known, offer either two or four uncompressed independent channels of digital audio. Some formats offer two additional FM or hi-fi tracks: here the audio is recorded as part of the picture signal and cannot be edited independently of the picture without being laid off to another tape in post production.

Format	Number of audio tracks	Format	Number of audio tracks
Beta SP	2 × linear channels, 16-bit @ 48 kHz 2 × FM	DVCPRO	4 × channels, 16-bit @ 48 kHz
DigiBeta	4 × channels, 20-bit @ 48 kHz	HD CAM	4 × channels, 20-bit @ 48 kHz
Beta Sx	4 × channels, 16-bit @ 48 kHz	DVCAM	2 × channels, 16-bit @ 48 kHz or 4 × channels, 12-bit @ 32 kHz

Figure 8.7 Typical acquisition formats and audio track specifications.

In terms of portability on location, it is easier to record sound using a camera with an integral or *dockable* recorder. However, in the studio, the output of a camera will be fed back to a free-standing VTR in the control room.

For many video applications, two channels of audio will be sufficient, either to record a stereo pair in the appropriate circumstances or to record two mono tracks of dialogue. This could consist of a boom mix on track 1 and a personal mic mix on track 2, or more simply, a single personal mic on track 1 and the on-camera mic on track 2. For some applications, four-track formats such as DigiBeta are more useful – for example, in a studio recording with audience participation, the stereo programme mix can be recorded on tracks 1 and 2, whilst the clean audience feed can be recorded in stereo on tracks 3 and 4.

Recording to multiple cameras or VTRs

Where multiple broadcast cameras are required to cover a scene or location from a number of angles, the output of each is fed to a vision mixer and the resulting programme mix recorded to a master videotape. However, for safety, the individual or *isolated* output of each camera is also recorded to its own tape. These tapes are known as the *ISOs*. Whilst the programme sound mix is sent to the master videotape, each individual mic output can be routed from the mixing desk back to the ISOs and recorded on its own channel. In a three-camera shoot using DigiBeta, for example, there are effectively 12 audio tracks available to the mixer (although two of them may be occupied by the programme mix).

Whilst it is faster to work with the mix in post production as far as is practicable, the sound editors may need to access the ISOs occasionally to replace a line clipped by a late fader move, or to deal with over-lapping speech. In a studio situation, each camera in a multiple set-up must receive *house sync*. On location, it is usual to use one of the cameras as a master, to which all other cameras are slaved or *gen locked*. Either method ensures that all devices are held in absolute sync, and avoids any drifting sync problems between sound and picture, which will be expensive and time-consuming to fix in post.

Recording to a separate audio-only recorder

Where a project is shot on film, the audio must be recorded separately to the pictures, using an audio-only recorder – this is known as the *double system* of shooting and requires an accurate method of maintaining sync between the camera and recorder. This technique may also be used on some video productions where the number of mic sources exceeds the number of record tracks available and mixing down is not an option, or where the sound recordist needs to operate independently of the camera to achieve better quality sound. On video shoots it is often a good idea to route the audio from the mixer to both the camera audio tracks *and* the audio recorder. This creates an automatic backup in case problems arise with the sound rushes, and also gives the picture editor a good-quality mix to cut with.

Recorders currently in professional use are summarized below.

DAT field recorders (Digital Audio Tape) are currently the most common type of location or *field recorder*. DAT machines can record two tracks of uncompressed audio at sample rates of 16-bit/44.1 kHz and 16-bit/48 kHz, together with SMPTE timecode. DAT tapes have a maximum of 2 hours running time, and ID markers or *programme numbers* can be added so that the sound editors can quickly locate the start of each take. Standard Digital I/O includes AES/EBU and S/PDIF, as well as analogue XLR connections. DAT is pretty near a universal medium and most tracklaying rooms and studios will have a DAT deck, which the editor can use to transfer recordings from the DAT tape to an audio workstation for editing. DAT recorders are fairly robust and will withstand motion whilst in

Figure 8.8 The HHB Portadrive: a file-based field recorder (courtesy of HHB).

Figure 8.9 HHB Portadrive: I/O configuration on back panel (courtesy of HHB).

record mode – this, in addition to its compactness, makes DAT a good medium for location work. DAT machines are now no longer manufactured as location recorders, but the large number of machines currently in professional use means that this format will still be around for a long time to come.

File-based field recorders are now being manufactured instead of DAT and will gradually replace it as the standard industry field recorder. Audio is recorded as a computer-compatible file onto a hard drive, DVD-RAM, DVD-RW or other removable storage device. Standard I/O configurations are also supplied – usually digital AES/EBU and analogue XLR. The recorders themselves are smaller than many DAT machines, and are built to withstand movement and rough handling whilst in record mode. Tape changing is completely eliminated, and the comparative lack of moving parts means that these recorders are not prone to mechanical failure (see Figure 8.8).

An important feature is the *pre-record buffer*, which allows the recorder to continuously store between 2 and 15 seconds of audio (individual recorders vary) when switched on. This ensures that the sound recordist never clips a take and could be particularly useful in news and documentary, where the recordist may have to react quickly. Recorders range from inexpensive two-track CompactFlash or hard disk field recorders, to high-end multitrack field recorders that can record between four and 10 audio tracks and support very high sample rates (see Figure 8.9).

Most high-end recorders have the ability to record simultaneously to two or more storage devices, so that a backup is instantly created whilst shooting. This is particularly important where picture and sound rushes are sent off at the end of the day for overnight processing, as there may not be enough time for the recordist to do a safety backup once the crew have wrapped for the day. Audio is recorded in file formats that can be read directly on a Mac or PC, and some systems can export audio as a Pro Tools or AES 31 session that can be imported directly into an NLE or DAW. Alternatively, the sound rushes can accompany the picture to the lab, where they will be synced together prior to viewing and editing. Audio can be recorded as a monophonic or polyphonic file, together with its associated metadata, such as roll/scene/slate number, circled takes, location markers, etc (see Chapter 4). The recorder also gives each recording a *unique identifier*, which all subsequent systems will use to locate the cue regardless of any cue name changes.

Model	No. of audio tracks	Storage medium	File format	Bit rate/sample rate
Marantz-PMD670	2	Compact Flash or Hitachi Microdrive	MP2, MP3, WAV and BWAV	Selectable bit rates/up to 48 kHz
Fostex FR2	2	PCMCIA 1.8-inch HD or CompactFlash	BWAV	Up to 24-bit/192 kHz
Fostex PD6	6	Mini DVD-RAM	BWAV	Up to 24-bit/96 kHz
HHB Portadrive	8	Removable Gig HD and DVD-RAM backup unit	BWAV or SD2	Up to 24-bit/96 kHz
Aaton Cantar X	8	Internal HD or Flash card and external HD	BWAV	Up to 24-bit/96 kHz
Zaxcom Deva X	10	Internal HD and DVD backup/external HD	BWAV, SD2 and ZAX file	Up to 24-bit/192 kHz

Figure 8.10 File-based recorders compared.

Digital audio workstations are sometimes used as multitrack recorders. DAWs are neither portable nor built for outside use, but until the recent introduction of multitrack field recorders, this was a belt and braces approach to recording in situations where a very high number of mic inputs needs to be recorded separately. Some TV projects such as soaps and 'reality' shows rely on a high number of personal mics to capture spontaneous and unchoreographed action. Under these circumstances it is safer to record a good separate track of each mic output, rather than trying to produce a usable mix. These separates can then be cleaned up in post production prior to mixing.

Modular digital multitrack recorders are designed to be rack mountable and mains powered, and as such are not particularly suited to location work. As portability is not an issue, these recorders come with enhanced editing features as well as eight or more record/playback tracks. If a higher track count is required, a number of units can be stacked together. Their main function in a broadcasting and film environment is as a *dubber*, onto which premixes and mixes are laid back.

Non-timecode recorders cannot be used for any project that involves sync sound, as an accurate lock to picture cannot be guaranteed and may result in drifting sync. However, where both picture and sound are recorded to videotape, it may be preferable to collect any non-sync material or wildtracks on a small portable recorder such as an inexpensive non-timecode DAT recorder or a MiniDisc recorder. This allows some flexibility and independence from the camera, and may be useful in documentary situations where unobtrusive recording is called for. Non-timecode DAT recorders offer the same recording facilities as their timecode counterparts, but do not have the ability to accept/record and replay timecode.

MiniDisc recorders are often used by sound effects editors to record custom fx in the field, and by documentary makers shooting on formats such as DVCAM, to record wildtrack fx, interviews and other non-sync sound. Recording takes place at a sample rate of 44.1 kHz, up to a bit depth of 16 bits. Audio can be recorded in mono or stereo via analogue or digital inputs. A buffer eliminates any interruptions due to movement whilst in record mode. MiniDisc recorders are extremely compact and one disc can hold about 2½ hours of mono material. In order to achieve this, the format uses ATRAC data compression, which reduces the data stored to one-fifth of its original full bandwidth. MiniDisc is therefore quite well suited to the recording of simple voices or single effects. It is not suitable for the recording of effects that will be later processed via samplers, for example, or for the recording of complex sounds (such as a mix), where the data reduction would be much greater.

Synchronizing sound and picture

Where sound and picture are recorded independently, sync must be maintained over a period of time. In order to do this, the camera and recorder must be *crystal controlled* – this accurate sync pulse system enables both to run precisely in real time without the need for any physical link between them. This means that when the sound and picture are eventually synced together in post, they will stay in sync. Non-crystal-controlled cameras cannot be used for projects that require accurate lipsync because minute fluctuations in the speed of the motor mean that the camera and recorder will gradually drift apart by a number of frames over a period of time. Crystal sync does not, however, provide any sync reference between sound and picture – it is merely a means of maintaining synchronization. There are essentially three methods of syncing film or video cameras to a sound recorder.

The clapperboard

This 65-year-old method of syncing sound and picture is still in use today. The roll/scene/take number should be written on the board, and the camera assistant should close the board firmly, whilst verbally announcing the slate and take number. In post production, the transient of the clap on the soundtrack can then be manually synced with the exact frame of picture on which the clap is seen to close. Where a shot is recorded as picture only, the slate should be marked as *mute* or *MOS* and a finger should be inserted between the arms of the clapper to indicate to post production that there will be no sound. Sometimes it is less disruptive to *ident* a shot at the end of the take rather than the start: this is known as an *end board*. This is indicated at the start of the shot by the camera assistant, who calls out 'end board'. Once the director has called 'cut', the shot is verbally idented in the usual way, but the board is held upside down whilst being clapped.

The digislate

This differs from the above in that the board has a large LED display which reproduces the timecode of the audio recorder. The board receives timecode via a cable or a wireless transmitter. In post, the syncing up is achieved by loading the audio (with timecode) into a workstation. The exact frame of picture on which the clap occurs is located by the operator and the timecode is punched into the system, which then searches for a matching audio timecode and automatically syncs up the shot – unfortunately, it is still partly a manual process.

Timecode

All video cameras in professional use can generate timecode or accept it from an external source. This will be recorded onto the tape along with the pictures, whether or not the audio is recorded separately. However, this is not the case with film cameras. Historically, many 16-mm cameras were timecode compatible as this format was often used for documentary filming where the use of a clapperboard was thought to be impractical and disruptive. On film sets, where clapperboards were a fact of every-day life, there was thought to be little advantage in shooting 35 mm with timecode. However, some 35-mm cameras are now available with timecode systems (e.g. Aatoncode). Where both picture and sound have been recorded with timecode, rushes can be instantly synced on-the-fly to the first frame of each take using a specialist DAW such as Aaton InDaw in what is a fully automated process.

Jamming sync

When a film camera (with timecode) or a video camera is physically linked to the recorder by a cable, sync problems are unlikely to occur because the two are locked to the same timecode. Unfortunately, this does mean that the sound recordist and camera operator are effectively tied together, and the flexibility of having a separate audio recorder is lost. Where the camera and recorder are not connected physically, even where both are crystal controlled and both start off from the same timecode numbers, there will still be some timecode drift between the devices over a day's shooting.

To minimize this drift, timecode is generated first by one device, and then *jam synced* into a second device a number of times a day. Jam sync means that the recorder, for example, synchronizes its internal timecode generator to match the start numbers from the camera, which may act as the master. To achieve this, an external sync box can be used – this may minimize drift down to as little as one frame over a day's shoot. The sync box is first attached to the camera's BNC timecode out connector, and the box accepts the camera's timecode. The box is then connected to the BNC timecode in con-nector of the recorder, which is forced to reset to the incoming code. Once the connection to the recorder is released, the recorder will stay in sync with the camera for a number of hours, at which point the process must be repeated. If this is not done, the sound editors will be faced with a drifting sync problem that will be very time-consuming to fix.

Most internal/external timecode generators have four settings that are used in conjunction with the jam sync function:

- *Free run/time of day*. This means that the internal timecode generator keeps clock time, record-ing the actual time of day. It runs continuously, whether the device is in record or not. This is a useful setting when taking notes of timecode numbers, as the sound or production assistant can

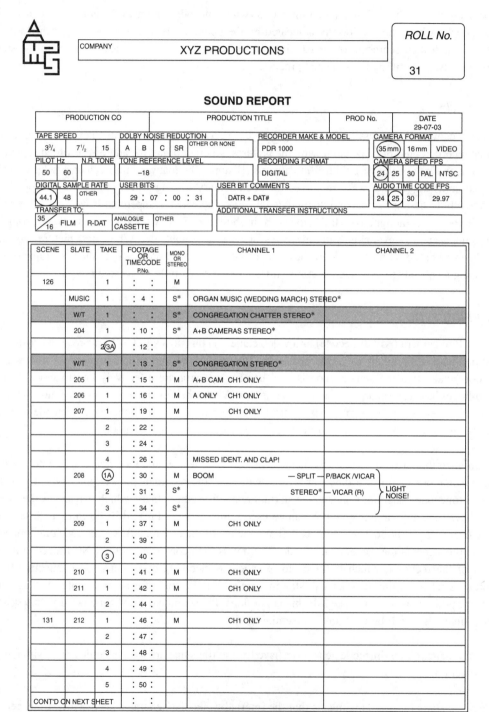

SCENE	SLATE	TAKE	FOOTAGE OR TIMECODE P.No.	MONO OR STEREO	CHANNEL 1	CHANNEL 2
126		1	: :	M		
	MUSIC	1	: 4 :	S*	ORGAN MUSIC (WEDDING MARCH) STEREO*	
	W/T	1	: :	S*	CONGREGATION CHATTER STEREO*	
	204	1	: 10 :	S*	A+B CAMERAS STEREO*	
		2 3A	: 12 :			
	W/T	1	: 13 :	S*	CONGREGATION STEREO*	
	205	1	: 15 :	M	A+B CAM CH1 ONLY	
	206	1	: 16 :	M	A ONLY CH1 ONLY	
	207	1	: 19 :	M	CH1 ONLY	
		2	: 22 :			
		3	: 24 :			
		4	: 26 :		MISSED IDENT. AND CLAP!	
	208	1A	: 30 :	M	BOOM — SPLIT — P/BACK /VICAR	
		2	: 31 :	S*	STEREO* — VICAR (R)	LIGHT NOISE!
		3	: 34 :	S*		
	209	1	: 37 :	M	CH1 ONLY	
		2	: 39 :			
		3	: 40 :			
	210	1	: 41 :	M	CH1 ONLY	
	211	1	: 42 :	M	CH1 ONLY	
		2	: 44 :			
131	212	1	: 46 :	M	CH1 ONLY	
		2	: 47 :			
		3	: 48 :			
		4	: 49 :			
		5	: 50 :			
CONT'D ON NEXT SHEET			: :			

PRINT CIRCLED TAKES ONLY : 2 CIRCLES = 2 PRINTS : A.F.S. = AFTER FALSE START

Figure 8.11 Sound report sheet (courtesy of Martin Trevis).

consult an ordinary watch, but results in discontinuous code, which may pose a problem in post production.

- *Free run/user setting*. This is similar to the above except that the start timecode is chosen by the user, rather than showing time of day. Sound roll numbers are often used as the first hour's reference (e.g. sound roll 6 starts at 06:00:00:00).
- *Record run*. Here, timecode is only generated in record mode and is suspended when recording is stopped or paused. The timecode restarts without interruption when recording is resumed, resulting in a continuous timecode track throughout the recording.
- *External run*. This means that the recorder reproduces timecode received from another device. If the external code is interrupted in any way, the timecode of the recorder will also be interrupted, although some recorders have an inbuilt system that jam syncs the timecode to the last good code and continues recording as though it were still receiving external code.

Identing and logging takes

Each sound roll/tape should be given a clear verbal ident confirming production title frame rate, sample rate, mono/stereo/split track and line-up tone level. This information should be repeated on the sound report and label for that roll. In order to help post production find their way around the shot material, it is important that all scene, slate and take numbers are logged on to a sound report sheet, together with any other information that may be useful. This may be whether a shot is a wide, medium or close-up, a brief description of its content, and perhaps a comment on sound quality (e.g. lighting buzz/plane overhead) – in fact, anything that may speed up the location of good takes. The preferred takes or *circled takes* are usually indicated on the report sheet, and can also be logged as data by some camcorders. Many report sheets are preprinted forms that are filed in by hand; however, logging programmes are available that may be run on a laptop and fed with radio-linked timecode from the camera. An example of a sound report sheet is shown in Figure 8.11.

File-based recorders such as Deva have the on-board capability to generate a sound report. The sound recordist can type in scene and slate number plus any comments, and the recorder will automatically create an entry for each take until the scene number is changed.

Studio-based recording

Crew

A typical television studio will have three intercommunicating control rooms: the first will be occupied by the lighting director and lighting board operator, the second by the director, vision mixer and producer, and the third by the *sound supervisor* (see Figure 8.12). The sound supervisor is responsible for mixing the studio sound and is assisted by the *grams op*, who will prepare and play in pre-recorded effects on cue during a take. On the studio floor, the *crew chief* (often a boom operator) will have planned which mics are to be used in which scenes and will supervise all sound rig operations during

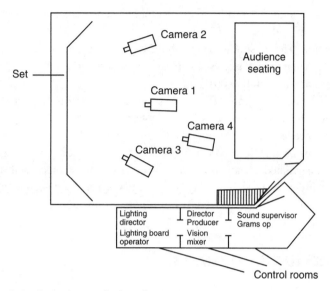

Figure 8.12 A typical studio and control room layout.

the recording. The crew chief may be assisted by *boom swingers* and *trackers*, who are responsible for moving Fisher booms around silently during taping.

Setting up

Studio-based programmes that will be finished in post (rather than broadcast 'live') can be approached in two ways. The programme can be shot 'live to tape'. Here, the supervisor combines all the required elements for the programme into a finished broadcast mix as though it were a live broadcast. The studio mic feeds, audience mic feeds, sound effects, music and any incoming OB/satellite feeds will be routed through a full-size studio mixing console equipped with multiple line and mic inputs. Alternatively, the programme can be recorded without sound effects, etc. – here, the intention will be to picture edit, tracklay and mix in post production.

In either case, the mix is routed to the audio tracks of the programme master, which will also be receiving the output of the vision mixer. For safety it is usual to simultaneously record the audio mix to a multitrack format such as DA88. When recording is complete – approximately 2 hours' worth of material may be shot for a 1-hour programme – the master programme tape and the ISOs (see above) can be given to post for editing.

Starting a take

The master VTR and ISOs are put into record mode and after a few seconds the tape operator calls out 'speed' – this indicates that all devices are up to speed. Colour bars and tone are recorded at the head of each tape, ensuring that the correct colour balance and audio level can be lined up on playback.

This is followed by an identifying slate and a countdown clock that goes to black at 2 seconds. After exactly 2 seconds of black, recording should begin. Recording is stopped by the director, who will say 'cut'.

Examples of studio-based recording

A *sitcom* set may typically be miked using three Fisher booms (depending on the size of studio) fitted with omnidirectional mics (such as the U87). These will offer wide coverage of the entire set. These may be augmented by a couple of directional mics on fishpoles that can be moved around the set as needed by the boom ops. Spot or planted mics may be concealed around the set as required. Depending on the size of the audience a number of slung stereo mics will be used – perhaps between eight and 12 for an audience of 500. The most important audio elements in the mix will be the dialogue and the audience track. The sound supervisor in the control room will mix the set mics with one hand and the audience mic mix through the other (this will be on a single stereo fader). He or she will compress the dialogue and push up the audience fader to anticipate the laughs in order to create as dynamic a mix as possible.

A *quiz show* set-up may use a number of line mics to cover the participants if they are likely to remain stationary, and the cables can be successfully hidden on set. Where the participants are likely to move around freely, radio mics should be used. The audience will be miked by a number of slung mics and a fishpole mic may be used to cover any specific contributions from the audience. In the control room, the supervisor will be balancing the set and audience mics as before, but will also incorporate sound effects triggered by the grams op and any VT inserts that may be required.

Delivering to post

Where a programme has been recorded live to tape, the master programme tape can be digitized into an on-line picture editing system, edited and output to the broadcast master tape very quickly, sometimes for broadcast on the same night as the recording. Because most of the sound post has already been done in the studio, it will only be necessary for the editor to perhaps add the odd effect, and occasionally access the ISOs when a dialogue overlap occurs. This technique is very useful for programmes such as quiz shows, sport and current affairs, where the soundtrack consists mainly of studio voices, audience reactions, some music and a few spot effects. However, formats such as sitcoms and children's TV are more suited to being tracklaid and mixed in post. In the UK, it is conventional to fully tracklay studio-based sitcoms and lay at least one stereo atmos per scene. This would be difficult to achieve under pressure in the studio, so a one-day dub may be booked to tracklay and mix within a single session. (This is not the case in North America, where sitcoms are left relatively clean of fx.)

Field/location recording

When recording on location, it is often difficult for the sound recordist to control sound quality because of external factors present in the recording environment, such as a nearby airport or motorway, as well as noise created by the filming process itself. It is important that sound crews anticipate and minimize the causes of noise, as background levels can, at the very least, make a recording sound very unpleasant, and at worst render it unusable. This issue is much more critical for certain types of production, where 'clean' dialogue is a given, and much less critical for factual television.

The main causes of 'noise' are as follows:

- *Lighting gear* is the most common cause of background hum, which can sometimes be removed by notch filtering in the mix. It is usually a 'buzz' (i.e. a square wave – lots of harmonics) and of high frequency, making it virtually impossible to remove with normal parametric equalizers, even digital ones.
- *Camera noise* is distinctive and difficult to filter. It is most noticeable in quiet scenes where the camera is close in on the actors. Camera dollies can be prone to squeaking during tracking shots. It may be possible to edit out the noise, or if it is under dialogue, use ADR to cover the camera move.
- *Set design*. Creaky floors and noisy props can affect the audibility of the dialogue and be intrusive. Carpet and foam rubber tape can reduce noise from crew and actors. Creaks/bangs between lines can be edited out to some extent and it may be possible to refit noisy takes with clean takes, if they are available. This is time-consuming and means that the director's chosen performance may be replaced by another take.
- *Noisy locations* are a problem if the background noise is inappropriate to the scene or if the dialogue is inaudible. If the level of the dialogue is only just acceptable, it may not leave many options in terms of adding sound fx when the scene comes to be tracklaid. Noisy but consistent backgrounds may be successfully equalized and gated in the mix. Plane noise recorded under dialogue can rarely be removed in post production. If the take is used it will be particularly noticeable when intercut with another shot. The affected lines should be considered for post sync.
- *Sound stage* acoustics often have a long reverb time – this may be triggered by projected or shouted dialogue, which will then sound appropriate to the size of the studio and not the set. This may be fixed by ADR or, if extensive, by fading off the reverb at the ends of words. Post sync any dialogue where the reverb overlaps. Dialogue spoken at normal level should be fine.
- *Crew noise* can be reduced by strict on-set discipline during a take. Noise from crew jackets and footwear can be a problem, particularly during camera moves. This should be cut out where possible, and covered with ADR if not.
- *Special fx equipment*. The use of rain/wind/smoke machines usually means that the dialogue needs to be post synced. Often, fire fx are created using gas, which may be audible under dialogue.

If these problems occur in a scene they can often vary from shot to shot, which means that if the original recording is to be used, each line has to be equalized and filtered separately in the mix, a time-consuming operation.

News/documentary

No sound crew

Most news teams do not include a sound recordist. This job is covered by the camera operator, who records audio back to the camera via a small portable mixer. A similar arrangement is often used in documentary work, where budgets are often too small to cover a sound crew. A typical arrangement would be to have the presenter/interviewer on a personal mic so that they are *always* on mic, regardless of where the camera is pointing. This leaves the on-camera mic to cover whatever the camera is looking at. Whilst this may not be so great for sound quality, it is simpler than constantly refitting a fragile radio mic to several interviewees. The on-camera mic is also more likely to pick up a number of voices than the personal mic. As sound and picture are recorded together, there is no need to use a clapperboard, although a board may still be used to identify each shot. Any wildtracks can be recorded separately to a non-timecode portable format such as MiniDisc.

With sound crew

Where a sound operator is included on the crew, he or she will be able to carry both the recorder and operate a boom, which means that the quality of sound should significantly improve with mic proximity. More sophisticated mic techniques become possible using a combination of the boom plus personal mics. These can be routed through a small production mixer (typically with four mic inputs) to the recorder, and sent as a backup recording to the audio tracks of the camera via a wireless transmitter. The sound recordist is also free to collect wildtracks that may be invaluable to the sound editors, particularly if the location or prop is difficult to gain access to (e.g. The Arctic, an icebreaking ship).

Drama/film

Shooting with a single camera

Most TV drama and feature-film location shoots are shot *discontinuously* using the *single camera* technique. The action is filmed, the camera is stopped and moved, the scene is relit, and the process repeated until all the required shots that may be required for a scene have been shot. A *scene* refers to the segment of a script that takes place in one location. A shot refers to a specific framed image within the scene and is given a slate number. Each filmed version of a shot is referred to as a take (e.g. Slate 267 – Take 3). A shot can simply be described as a wide shot (WS) or master shot, a medium shot (MS), or a close-up (CU). These terms can also be applied to the sound and its perspective. Although the shots will eventually be edited together into a complete scene, they are usually shot not in the order they will eventually appear in the cut, but in the order that involves the fewest camera moves on set.

Crew

When the production mixer has been hired for a production, he or she will recce (or scout) the chosen locations with other members of the crew and decide how best to mic each scene. For example, the director may be consulted about frame sizes and whether tracking shots are to be used, and the director of photography could indicate how a scene is to be lit, which may affect the use of the boom (and cause the boom's shadow to fall across the shot). The production designer and costume designer may be invaluable in concealing planted and personal mics. The production mixer will be assisted by one or more boom swingers and/or a sound assistant, who may double as a second boom swinger when needed, in addition to keeping track of equipment and paperwork. (In the USA, these crew members are known by the alternative titles of boom operator, cable person and cable puller respectively.)

Starting a shot

There is a set procedure for starting a shot – this is particularly true of film shoots, where the raw film stock is very expensive and wastage kept to a minimum. The director instructs the sound recordist to 'turn over'. Once the recorder is up to speed, the recordist confirms 'sound speed'. The director then calls 'roll camera' and this is confirmed verbally by the camera operator. At this point, both the camera and recorder are running in sync with each other, and the director can call 'mark it'. The camera assistant or clapper/loader announces the slate and take number, and claps the board, at which point the director says 'action'. The take is stopped only when the director says 'cut'. On the occasions where a second camera is used, each camera should be clapperboarded separately with its own board. Boards should be marked and verbally idented as Camera A and Camera B, so that it is clear to post production which sync mark goes with which camera.

Wildtracks

Any audio recorded on location independently of the camera is known as a *wildtrack*. Dialogue wildtracks are often recorded when the recordist knows there will be a problem with the dialogue spoken during a take – for example, a loud car door over a line, a noisy location or a dialogue overlap. Many recordists will record a clean *buzz track* of the atmosphere or scene ambience, which may be useful to the editors when smoothing out the production sound (see Chapter 11). Fx wildtracks can be invaluable to sound editors, particularly where a production involves a particular prop, crowd or location that may be difficult to cover with library fx. A specific model of motorbike, for example, should be recorded idling, approaching and stopping, starting and leaving, and passing at different speeds. Using these tracks, the sound editor will probably be able to cut fx for the motorbike as it appears in each shot. Each wildtrack should be verbally idented by the recordist and clearly marked on the sound report sheet. In practice, many productions are shot on such tight schedules that there simply isn't time for the recordist to take a wildtrack without holding up the shoot.

Examples of discontinuous shooting

City restaurant/two actors sat at table

The traffic should be halted if possible and aircon switched off. Extras (supporting artists) should be asked to mime background chat to keep principal dialogue clean. Principal actors must keep their level up to a realistic level, as crowd tracks will be added in post production. Fix any noisy props (e.g. tape the bottoms of cups). A *two shot* (both the actors in shot) will be shot first, with one boom favouring each artist in turn. The boom operator needs to be familiar with the dialogue to anticipate changes in mic position. *Singles* will then be shot of each actor, with the boom favouring the actor in shot. If two booms are available, both actors should be on mic even though one is out of shot. This will allow editing even if dialogue overlaps. At the end of the scene, record a wildtrack of restaurant chat with the extras – this can be used in the tracklay later.

Noisy street/two actors shot on long lens

In this situation it will be impossible to get the boom close enough to the actors without getting it in shot. The recordist will need to use one radio for each actor, and these should be routed to separate tracks on the recorder. Apply as little eq as possible, as this can be done more effectively in post production. A multitrack recorder would allow the recordist to use a third and fourth track to record the stereo foreground traffic. Alternatively, if only two tracks are available, the recordist could mix both radio mics down to one track and put mono traffic on track 2, but a better solution would be to leave the radio mics across two tracks and wildtrack the traffic. If the location is so noisy that the scene may have to be ADR'd, a good guide track must be recorded, and it would be a good idea to take the actors to a quieter part of the location and ask them to wildtrack the dialogue.

Delivering to post

News tapes will be taken straight from the camera to the newsroom and digitized onto the server of a news editing system, or edited directly from the master in a linear tape suite.

Documentary material may be cloned as a safety precaution before being handed over to the editor, together with any sound report sheets and the separate audio masters (if these were recorded).

Drama/film material shot using the double system must be sent at the end of the day to be synced up overnight for viewing the following day – hence the name *rushes* or *dailies*. Timing is crucial in film shoots as the unprocessed film rushes must be developed, printed, assembled into rolls and *telecined* to Beta SP worktapes before syncing up can take place. There are many ways of syncing up, some of which have already been mentioned in this chapter. In the USA, syncing up tends to be done in telecine, although this is a more expensive option, and is relatively slow. There are actually fewer telecine machines available in the UK, and so syncing up is done after telecine. Audio rushes are imported from the location DATs, DVDs or hard drive into an audio workstation, which is connected to a Beta SP machine. Using an automatic or manual syncing technique, the sound for each shot is located and matched to the Beta SP using either timecode or the clapperboard.

For a *broadcast project*, the audio is then laid onto the Beta SP, and simultaneously recorded to a new DAT (one for each telecine roll) with code which matches the continuous code of the Beta SP. A *shot log* can be generated from the telecine session, which can be used by the editor to batch digitize material into the picture editing system. Once picture editing has been completed, the sound editors can replace the Beta SP sound used in the Avid by auto-conforming with the better quality DATs. As these DATs have timecode that matches the Beta SP tapes, the conform should exactly replicate the original audio edits in the picture edit system.

For a *film project* running at 24 fps, it is necessary to telecine at PAL video speed, which is 25 fps. This means that the picture on the Beta SP telecine is now running 4 per cent fast. Audio rushes are synced as for broadcast, but must also be speeded up by 4 per cent before being laid back to the telecined pictures. Once the Beta SPs are digitized into a picture editing system running at 24 fps, the original speed of both picture and sound is restored within the system. However, in order to relate the picture and sound back to the original camera rolls at the end of the edit, a *flexfile* is produced. This is effectively a database that can be used to trace any clip which exists in the picture editing system back to its original source. The flexfile can be used as a shot log to batch digitize material into the picture edit system. Once the picture edit is complete, a 24 fps audio EDL is generated and the *original* location audio (which is, of course, still at 24 fps) is used to exactly replicate the original audio edits made in the picture edit system.

There are other ways of working, including the option of syncing rushes up within the picture editing system itself. The editing process is looked at in more detail in the next chapter.

9 Editing picture and sound

Hilary Wyatt

An overview

This chapter follows the path of the location sound through the picture editing process. Decisions made at this stage will significantly affect the work of the sound editors further down the line.

In the last 10 years, computer-based editing has almost completely replaced tape and film editing in professional applications. Editors still choose to cut 35-mm features on film occasionally, usually for creative reasons. Tape-to-tape or *linear* editing is still used in a limited way in low-budget TV work as well as corporate video, TV promos and some news operations, where a fast turnaround can be achieved by editing tapes straight from the camera. Tape edit suites are also used to conform material that has been shot on tape (usually a digital format such as DigiBeta or HD) and edited in an off-line non-linear system (NLE). Tape editing is essentially a mechanical process which involves the transfer of selected material from one or more source VTRs to a record VTR. The VTRs are synchronized by an edit controller, which locks the machines together (via RS422 Sony nine-pin protocol) using control track or timecode. To perform a simple edit, in and out source points are entered into the controller, as well as an assembly in point. The edit controller then synchronizes both the source and record machine together during pre-roll, dropping into record at the selected in point and out of record at the selected source out point. The process is then repeated, editing the next shot on to the end of the previous shot until the assembly is complete, hence 'linear editing'.

Most tape formats have two or four linear audio tracks available and audio is generally checkerboarded across the tracks, with overlaps enabling smoother transitions between cuts. If further work needs to be carried out on the audio, the audio tracks may be laid off to a DAW in real time, or an audio EDL may be created which will be used to auto-conform the tracks into a DAW from the original source tapes. Alternatively, in news for example, all interviews and sync fx may be laid down to track 2 before being re-recorded or *bounced* to track 1 with commentary. Track 1 then becomes the broadcast (mono) master.

Figure 9.1 The timeline of a non-linear editing system (courtesy of Avid).

The main limitation of tape editing is its lack of flexibility. Unless the length of the new material is exactly the same as that which is to be removed, then changes can only be made by dubbing the entire assembly (with the new edit) to another tape. Where analogue tape is used, such as Beta SP, each tape generation used will mean a gradual loss in picture and sound quality, as well as the usual analogue problems of tape hiss and drop-outs.

Non-linear editing

The term *non-linear editing* is used to describe an editing technique that is almost entirely digital. All source media is first *logged* and then *digitized* roll by roll, into the NLE (non-linear editor), where each shot is held on a series of hard drives as a *clip*. When all the source media has been digitized and sorted into *bins*, the editor can select a clip and load it into the source monitor. The clip is played, and in and out points are marked. The clip is then edited into position on the *timeline*, and the process repeated with the next clip. The timeline is a graphical representation of the edit assembly, showing the video track positioned above the audio tracks.

On playback, each edited video or audio clip merely refers the system back to the master clip held on the hard drive and the process is entirely non-destructive. Material can be accessed instantly by clicking on to the selected point within a source clip or the assembly, eliminating the need to shuttle through tapes in real time. All tracks can be independently edited, material can be overlaid, inserted or removed, and different versions of the cut saved quickly and easily. Edits can be further refined using the *trim* function, and video/audio effects added. Some effects which cannot be played in real time will need to be *rendered*, producing a new clip. This becomes an issue for audio when plug-ins are used (see below). Once the project is complete, the edit can be outputted to tape or transferred to another system via OMF or an EDL.

System configuration

Most systems consist of one or two computer monitors, which display the software menus, source bins, timeline, and the source and assembly windows in which the rushes and the edit can be viewed.

The core of the system is a PC or Mac computer on which the operating software runs, such as Avid or Final Cut Pro. The internal hard drive is used to store all project data and saved edits. The actual media (picture and audio files) is stored on an array of external drives. Alternatively, multi-user systems such as Avid 'Newscutter' store media on a shared storage device, which can be accessed by many workstations via a high-speed network. A system such as Avid 'Unity' can store up to 5 terabytes of media on an array of drives set up in a RAID configuration, and media is sent and retrieved via a network (LAN or SAN) using both standard Ethernet and Fibre Channel interconnections (see Chapter 1, 'Filesharing and networking' section).

NLEs are supplied with a standard pair of speakers and a small mixing desk on which to monitor the audio output channels. Editing is usually carried out using a keyboard and mouse. In addition to the NLE, most cutting rooms will be equipped with a VTR (usually Beta SP), which can be used for making digital cuts, and other equipment such as a Jazz drive, VHS, DAT machine and CD player.

Video resolution

When first introduced, NLE systems were purely *off-line* machines which digitized material in at very low quality. It was always necessary to produce an EDL that would be used to conform the edit using

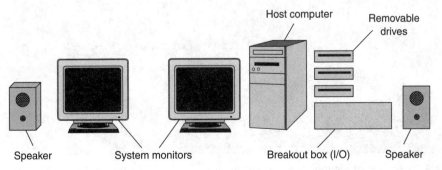

Figure 9.2 Typical non-linear editing system hardware.

the original film or tape rushes, which were of much higher resolution. However, higher processing speeds and the availability of larger hard drives mean that on-line quality pictures can now be outputted from some systems uncompressed. (Final Cut Pro and Avid DS, for example, can digitize both high definition (HD) and standard definition (SD) formats uncompressed.) However, for most projects some degree of picture compression is needed to reduce the drive space to a manageable size. For example, 1 gigabyte of memory will store approximately 3 minutes of video digitized at the high-quality video resolution AVR 77 (compression ratio 2:1). This setting would be suitable for short projects such as commercials and promos, but on-line systems are also used in many areas of TV production where a fast turnaround is required. News, sport and programme inserts are often edited and mastered directly to tape. An on-line system can also be used to redigitize the cut of a project edited on an off-line system at a higher resolution.

For long-form projects such as drama, features and documentaries with high cutting ratios, it may be necessary to use a low-resolution setting. One gigabyte of memory will store approximately 40 minutes of video digitized as an off-line project at AVR 3. The disadvantage of this is that any work tapes generated by the NLE will be of low quality with visible artefacts. This can have an implication for subsequent ADR and foley sessions, as some picture detail may be unclear.

At the start of a project, a video resolution is chosen and all subsequent material is digitized at that compression ratio. Sound quality is not affected by this setting because, unlike picture, audio does not require huge amounts of memory and is therefore digitized uncompressed. This means that if it is

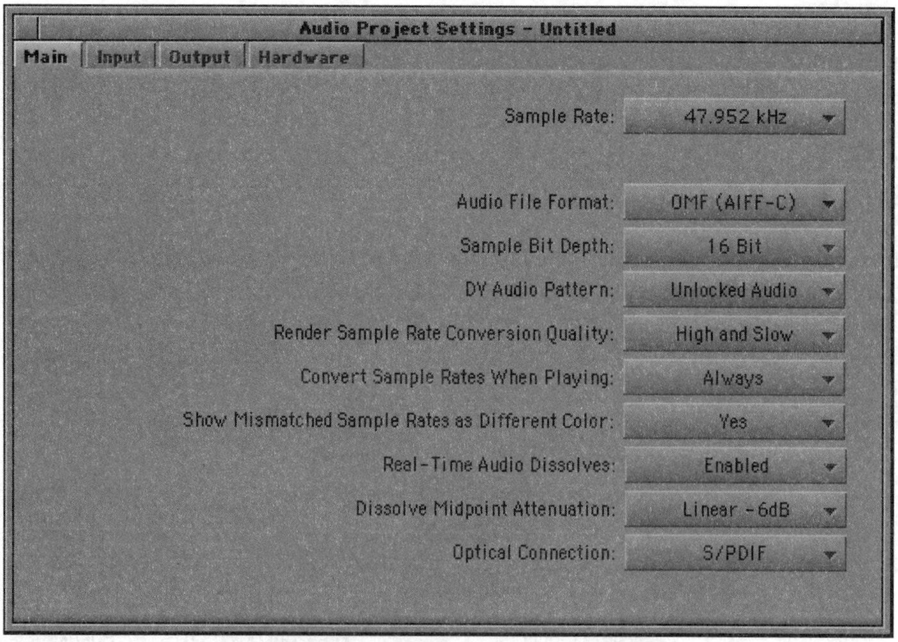

Figure 9.3 Audio project settings (courtesy of Avid).

loaded in at the correct level, preferably digitally, the audio can be used straight from the NLE rather than needing to be reconformed from the original location recordings at a later stage.

The editing process

Project settings for audio

When starting a new project, there are a number of important audio settings to be selected prior to digitizing.

File format

There will be an option within the NLE to set a default file format for the project. This will usually be a choice between AIFF and BWAV. If the project is to be exported to a sound facility, it is best to consult them as to which format to use. Failing this, BWAV is probably a better choice, as it is now supported by PC- and Mac-based editing systems, has time stamp capability, and is the designated native file format for AAF.

Sample rate

This setting controls the rate at which audio is played back in the timeline, and for all professional applications a sample rate of 44.1 or 48 kHz is used. Sample rates higher than 48 kHz are not currently supported by picture editing systems. DV camcorders can record audio at 32 kHz and some NLEs offer this additional setting. Audio recorded through the analogue inputs can be digitized at either 44.1 or 48 kHz, although the latter is slightly higher in quality. If the project is to be exported to a sound facility, it is important to check what their working sample rate is – most now work at 48 kHz as this complies with broadcast specifications. Where audio is digitized through the digital inputs, the NLE setting should match the sampling frequency of the original recording to avoid any loss of quality. Audio can be sample rate converted, so that a CD playing at 44.1 kHz, for example, can be digitized in at 48 kHz. Newer NLEs can convert non-standard audio in real time during playback.

Audio levels

The input level setting on the NLE is only critical when the audio is being digitized through the analogue inputs. In this case, the sound recordist's tone should be used to line up the meters displayed in the audio settings of the NLE. (Line-up tone is usually set at −18 dBfs in Europe and −20 dBfs in North America.) In the absence of tone, it will be necessary to shuttle through the recording to find the loudest point and check that it is peaking at just below 0 dBfs.

It is not necessary to set a level for audio that is digitized through the digital inputs, as the original recording level will be reproduced unaltered. This also applies to audio that is imported or *ingested* as a file, such as BWAV rushes from a file-based field recorder.

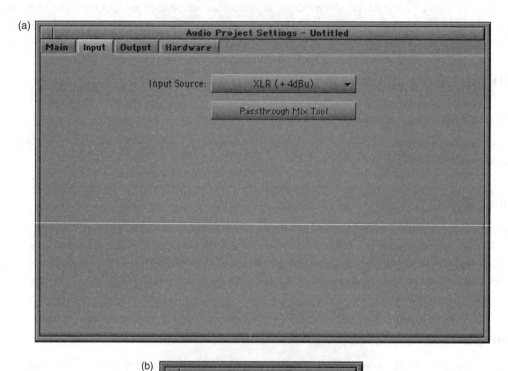

Figure 9.4 Audio input level settings (courtesy of Avid).

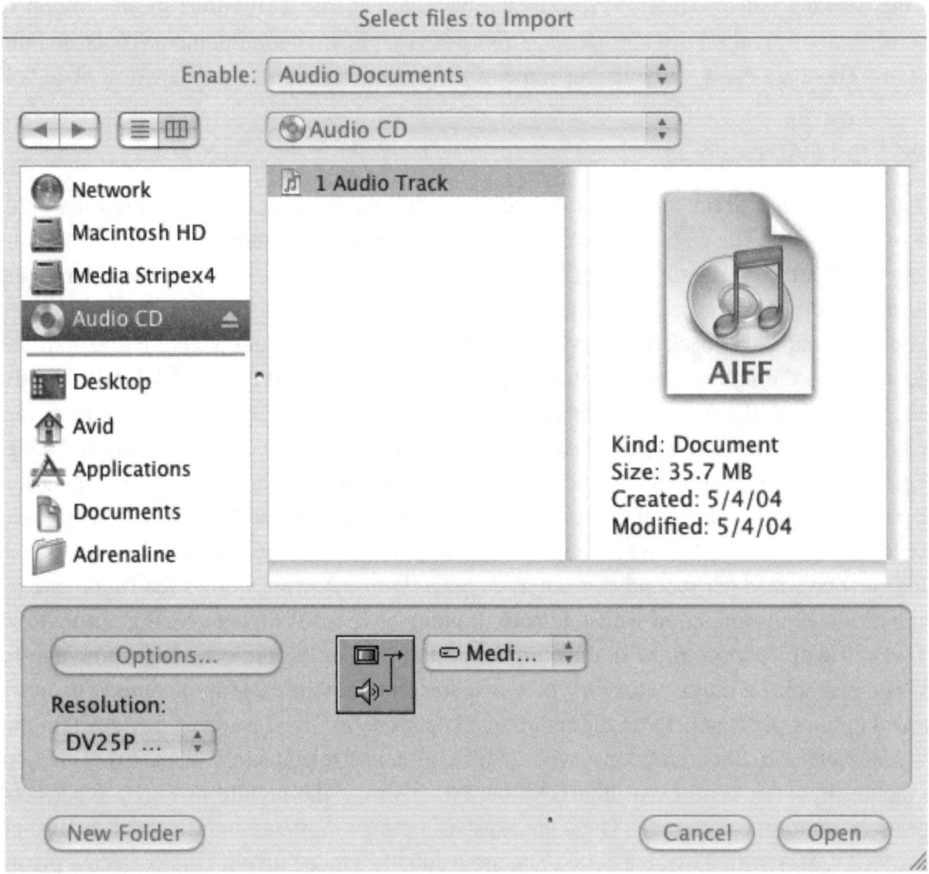

Figure 9.5 Importing audio as a file (courtesy of Avid).

Frame rate

It is important to understand the subtle difference between picture frame rate (the speed at which the pictures play out) and timecode frame rate (the numerical units in a second) – they are often but not always the same. Where a project is shot on film and then transferred to video before being digitized into an NLE, workflow methods have been developed which allow for the fact that the frame rate of the acquisition medium (film) is different to that of the editing medium (video).

- *Video*. The standard video frame rate for PAL is 25 fps, so the NLE should be set to 25 fps throughout. The video frame rate for NTSC is 29.97 fps, so the NLE should be set to 29.97 fps, but it is important to know whether source tapes are *drop frame* or *non-drop*. Most projects are post produced using non-drop as the standard. Drop frame is used in the making of broadcast masters, where the actual clock time of a project is critical.
- *Film*. The international standard for film playback is 24 fps. Now that film is edited using video technology, various methods have evolved to compensate for the speed difference between 24 fps

and the working video standard. Projects can be edited at true 24 fps film speed only on specific systems such as Avid's Film Composer, Lightworks or FCP running CinemaTools or FilmLogic plug-ins. There are many variants, but some of the most common methods are described below.

24 fps/25 fps PAL projects

1. *PAL film method 1.* Shoot at 24 fps (with 25 fps code) and telecine at 25 fps, speeding up the picture by 4 per cent. The audio rushes are also speeded up by 4 per cent and synced to the telecine tapes, which are now running fast at 25 fps. A *flexfile* is generated in order to cross-reference the telecine tapes back to the original timecodes. Once the rushes have been digitized, the NLE 'slows' both picture and sound back down to 24 fps so that original speed and pitch are restored. The NLE is set to 24 fps throughout and, on completion, a 24 fps sound and picture EDL will be produced. Using this method, the audio has to be conformed from the original location recordings, using information contained in the flexfile to locate each original clip. The overall sound speed change does not guarantee frame lock between picture and sound, and so the conform will only be accurate to ±2 frames. It is important to note that the project is only playing at 24 fps *whilst it is in the NLE*. In order to display the cut on a PAL video monitor, or record a playout to a PAL VTR, an extra two fields must be added per second (known as *24 fps pull-down*), which slows the frame rate by 4 per cent back down to film speed whilst actually running at 25 fps. When an NLE is set up to work in this way, it will slow down *all* digitized material throughout the project unless it is ingested as a file. For example, a music track or voice-over recorded against 24 fps pictures will need to be speeded up by 4 per cent prior to digitizing to compensate for the 4 per cent slowdown in the NLE.
2. *PAL film method 2.* Shoot at 24 fps (with 25 fps code), and telecine at PAL 25 fps. This *speeds up* the picture by 4 per cent. Once digitized, the NLE 'slows' the picture down by 4 per cent, so it appears as it did in the camera. The sound rushes are now digitized into the NLE at 100 per cent film speed (24 fps with 25-frame code), and are manually synced up with the corrected picture. The NLE should be set to 24 fps throughout the edit and on completion, both the picture and sound EDL will be 24 fps. The advantage of this method is that the editor can work with full quality audio, which can be outputted directly as an OMF and used for the mix, completely eliminating the need to auto-conform back from the original tapes. As with the first method, any subsequent playouts must be made with 24 fps pull-down.

24 fps/29.97 NTSC projects

1. Shoot the film at 24 fps and sound with 30 fps timecode. Telecine the 24 fps picture adding 12 fields per second in a 2-3-2-3 sequence. This is known as *2-3 pull-down* and it effectively makes the video run at the same speed as the original film. Each 24 frames of film are now represented by 30 frames of video.This is, however, 0.1 per cent faster than the running speed of an NTSC video deck, so the film is slowed down by 0.1 per cent to run at 23.976 fps in telecine to produce tapes that play at 29.97 fps. The 30 fps audio rushes are also slowed or *pulled down* by 0.1 per cent to run at 29.97 fps. The audio can then be laid back to the telecined pictures and will sync up perfectly. This is achieved by adjusting the playback sample rate from 48 to 47.952 kHz, or from 44.1 to 41.056 kHz. Both sound and picture are digitized together into the NLE, which discards the extra fields introduced in telecine – this is called *reverse telecine*. The NLE will now play back picture and sound at

23.976 fps. (Confusingly, this is sometimes referred to as 24 fps!) The film is tracklaid and dubbed at this pulled-down rate, and is pulled up by 0.1 per cent to 24 fps at the mastering stage.

2. Shoot picture at 24 fps and sound at 30 fps. Telecine the film at 23.976 fps picture with 2–3 pull-down. Digitize the 29.97 fps pictures into the NLE with reverse telecine. Digitize the original location sound in with 0.1 per cent pull-down and sync to picture. The NLE will now play back picture and sound at 23.976 fps and the project will remain at this rate until the mastering stage, when it will be pulled up to 24 fps. Some NLE systems have a setting which *pulls up* the digitized picture to 24 fps, so that all editing is carried out at true film speed. In this case the original location sound should be digitized into the system at its original speed of 30 fps, with no pull-down. No further speed changes are necessary, and the project will remain at this speed through to projection.

Logging the rushes

Tape rushes are usually logged manually, often by an assistant editor. At its simplest, this entails logging the start and end timecodes, slate and take number, and any comments, which may include a short

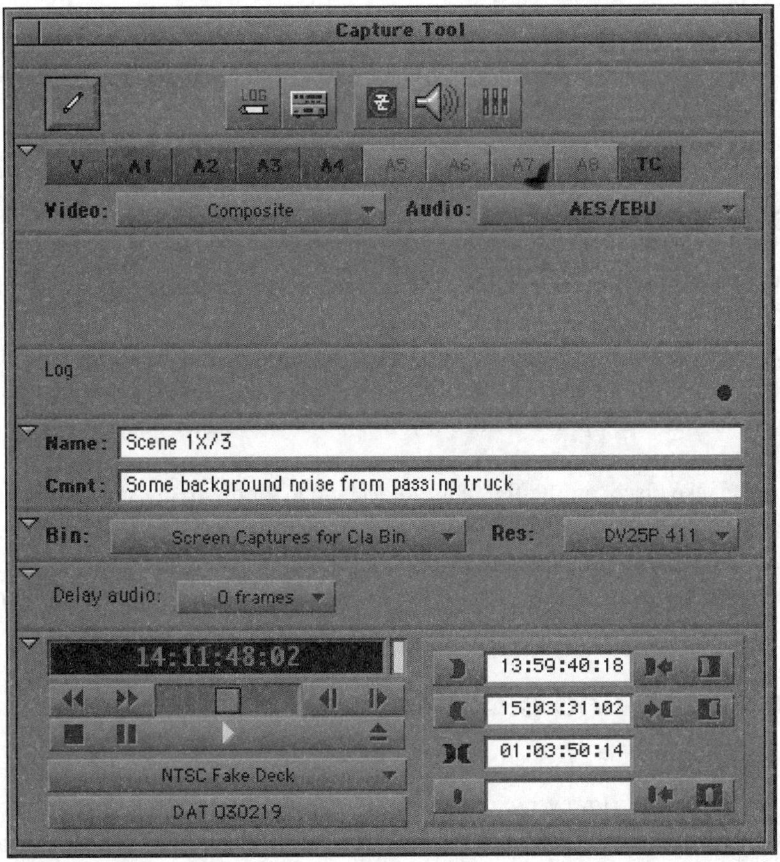

Figure 9.6 Rushes logged using the digitizing tool (courtesy of Avid).

description of the shot. It is vital to log all data correctly, as this information will be used to link each digitized clip back to its source later on. Logging can be done on a system such as Avid's 'Medialog', which is essentially a free-standing PC or Mac that drives a VTR via a nine-pin remote. Alternatively, logging can be done in the digitizing/capture window of the editing system. Shuttling through each tape, in and out marks are logged for each clip and the video/audio tracks to be digitized are selected. When the log is complete, the data can be sorted into bins and exported via floppy disk to the NLE prior to digitizing.

Film rushes that have been telecined and sound synced prior to delivery to the cutting room will be accompanied by a floppy disk containing the shot log or *flexfile* for each tape. This contains all the information needed to batch digitize picture and sound from the telecine tapes into the NLE. Flexfiles cannot always be read directly, so the file may need to be imported into third-party software, which translates it into a format that the NLE can read. Final Cut Pro, for example, uses a plug-in called 'FilmLogic' to open flexfiles; Avid users need to import the file into 'Avid Log Exchange', which produces an .ALE log. This file will automatically drive the NLE and VTR to locate and digitize each logged shot, which is then saved as a digital clip.

If it is decided to sync rushes in the NLE itself, the shot log is first used to batch digitize pictures only from the telecine tapes and then to batch digitize the audio from the *original* sound rolls, which may be on DAT. Rushes that arrive at the cutting room as BWAV files are *ingested* into the NLE in a matter of seconds per take, eliminating the need to digitize in real time.

Digitizing sound and picture

Digitizing takes place in real time, and this can be a lengthy process. The fastest way to handle large quantities of rushes is to *batch capture*, or *batch digitize*. Open the bin containing the logged clips, select the clips to be digitized, and open the batch digitize window. The NLE will drive the video/DAT deck to the start of each logged clip and automatically drop in and out of record before moving on to the next clip. At the end of the process, the bin will now contain *master clips*.

Digitizing can also take place 'on-the-fly', a useful technique for digitizing smaller quantities of material. With the video/DAT deck in play mode, the digitize button is pressed to start record and pressed again to stop. A new master clip will then appear in the bin. Alternatively, in and out marks can be set, and the digitize button clicked. The deck will locate to the in point and drop in and out of record.

Syncing sound and picture

Digitizing picture and sound together off tape creates a master clip in which picture and associated sound are linked as one digital file. However, on film projects where picture and sound rushes have been digitized separately, the two clips will have to be synchronized together before any editing can take place. This is done manually in the NLE using the clapperboard as the sync reference, and is known as 'syncing up'.

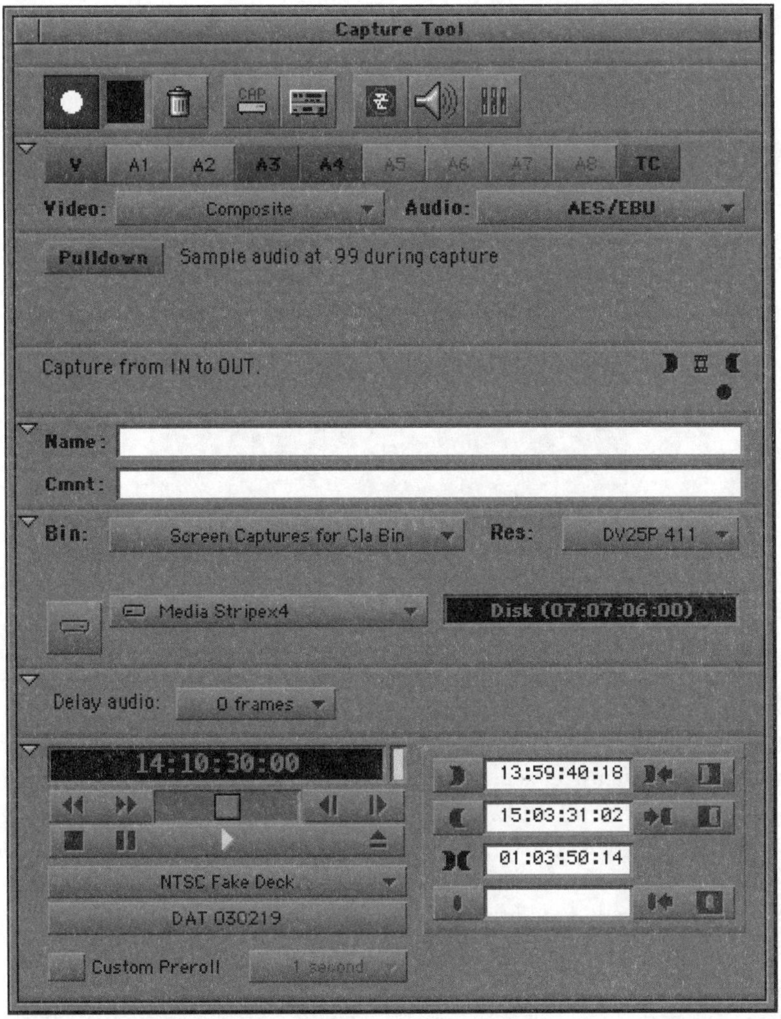

Figure 9.7 The digitizing tool (courtesy of Avid).

The picture clip is analysed to find the frame on which the clapperboard is just about to close, and a mark made at that point. The audio clip is analysed to find the start of the clap and a mark made at that point. The two marks are snapped together and a new master clip will appear. This clip will then be used in the edit.

Tracklaying and mixing within the NLE

The primary function of an NLE is picture editing rather than sound editing, and most types of project will be handed over to the sound department to be finished on a DAW. The standard speakers supplied with most editing systems are quite poor, and NLEs generally occupy rooms that are acoustically unsuitable for any kind of reliable audio monitoring. Systems are usually set up with the drives and associated hardware sitting in the same room as the editor, which means that all work takes place

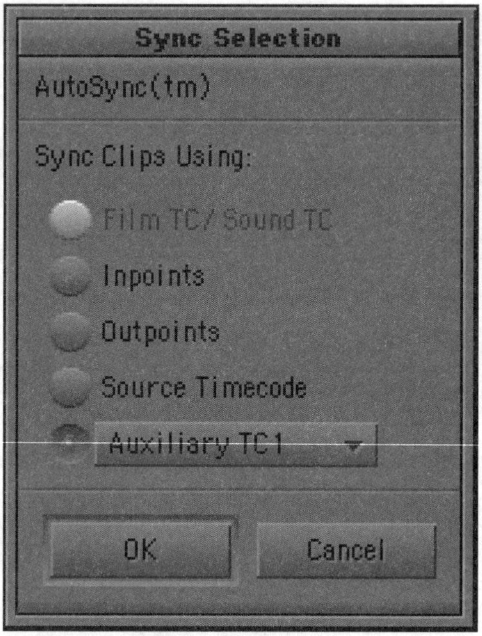

Figure 9.8 Autosync tool (courtesy of Avid).

against a high level of ambient noise. Despite these considerations, the audio editing functions on current NLEs are now fairly extensive, and for some applications such as news, documentary and sport, tracklaying and mixing within the NLE, and directly outputting to tape, may be the most cost- and time-effective option. At the other end of the spectrum, feature editors may lay up multiple temp music and fx tracks that can be balanced against the dialogues and outputted as a rough mix for viewing tapes. Most NLEs can generally support up to eight video tracks and eight audio tracks, which can be monitored simultaneously in the timeline. If this isn't enough, tracks can be freed up by creating sub-mixes of existing tracks using the on-board mixer, and all systems offer a selection of clip-based and real-time audio processing. (Final Cut Pro currently supports up to a maximum of 99 tracks going into 24 output channels.) However, one drawback of tracklaying on an NLE is that in order to preserve the frame-for-frame relationship between the digitized clips and source material, the system must not allow sub-frame edits. This means that an audio edit can only be positioned to the nearest frame line, which may not be accurate enough in some circumstances. (This is beginning to change: some newer NLEs such as Final Cut Pro 4 do have the ability to create sub-frame edits.) NLEs also only offer linear fades and symmetrical crossfades. Mixes can only be completed in mono or stereo – mixing in surround is not currently supported.

Editing audio in the timeline

A sequence is assembled by dragging master clips into the source window. In and out points are marked on the source clip and an in point selected on the timeline. Destination tracks can be chosen by clicking

Figure 9.9 Punch-in tool (courtesy of Avid).

on the track selector. Click on the add edit button and the new clip will appear in the timeline. The finished assembly will consist of one or more video tracks, with the production sound usually laid on audio tracks 1–4. Fine cutting can now begin, and audio edits can be refined by using the *audio scrub* tool, moving edit points, adjusting gain, and adding fades and dissolves between clips. Wildtracks and fx may be added, perhaps to audio tracks 5 and 6, and music may be laid using tracks 7 and 8. If the project is to be exported to a DAW for further work, this is probably as far as the editor will go in refining the audio edit, as any further audio processing will not be carried over by the EDL or OMF. (This is not true of some systems such as Avid and Pro Tools, which share the same core plug-ins.)

If the audio is to be finished within the NLE, the editor will probably spend some time adjusting levels and moving tracks around to smooth out the production sound. Further processing will then take place using the on-board audio tools.

Audio tools

● *Audio mix window*. Each track is assigned a fader, which controls clip-based level and panning. Channels can be grouped together so that any change in setting is applied across a number of channels.

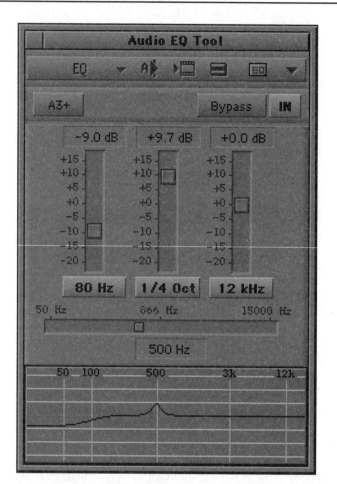

Figure 9.10 Audio eq tool (courtesy of Avid).

- *Audio punch-in.* This feature allows controlled recording directly to the timeline. The editor can set in and out points, and the NLE will automatically drop into record on a selected track. This can be used to record from any external device, but is particularly useful for recording voice-over to picture.
- *Gain automation.* Some NLEs offer gain automation, which enables the operator to ride the levels along a track using the mouse to make adjustments on-the-fly.
- *Audio equalization.* Some NLEs have a clip-based three-band eq window in which selected frequencies can be boosted or cut. The three faders (LF/mid-range/HF) are adjusted using the mouse, and the clip can be auditioned until the desired setting has been achieved. Most NLEs have a menu containing filter and eq presets, such as phone eq.
- *Plug-ins.* NLEs are supplied with a number of plug-ins as standard, many of which duplicate the functions of processors found in a mixing environment. The plug-in menu may contain settings for reverb, timeflex, pitch shift, compressor, limiter, expander and de-esser. The processed clip may be auditioned before it is rendered.

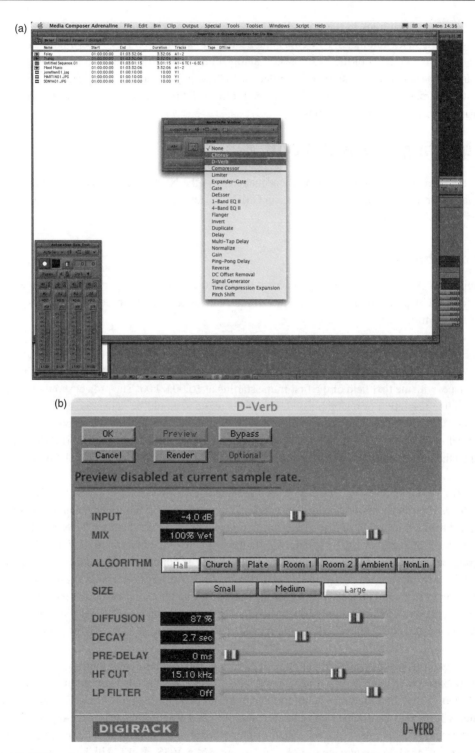

Figure 9.11 Plug-ins menu (courtesy of Avid).

- *Audio mixer*. Each track in the project will have an assigned channel strip on the mixer. The strip consists of a fader and meter, as well as mute solo and panning controls. A separate fader controls the master output level of the mixer. Mixer automation allows adjustments in level and panning to be made on-the-fly using the mouse. On playback, these adjustments will be accurately reproduced by the automation. This feature enables the editor to work through the sequence on a track-by-track basis until a balanced mix is achieved.

Outputting the audio edit

There are currently three ways to export a sequence from the NLE: direct to tape (sometimes known as a playout or digital cut), as an EDL or as an OMF. The advantage of one method over another is discussed in detail in Chapter 4.

Direct output to tape

As well as outputting on-line quality broadcast masters, the NLE can output viewing tapes which are often required by production, and work tapes will need to be generated for the sound editors, the composer and the mix.

Broadcast masters will need an ident, colour bars and a 30-minute countdown clock placed at the head of the sequence in accordance with delivery requirements. The sequence should be laid down to a striped tape, with the first field of the first frame starting at 10:00:00:00. The project should be exported at the on-line resolution or uncompressed. Broadcast masters should be delivered at 48 kHz and PAL 25 fps or NTSC 29.97 fps drop frame. 1 kHz line-up tone should accompany the colour bars at a reference level of PPM4, with programme levels peaking at but *not* exceeding PPM6. (PPM4 equates to 0 dBu on the analogue outputs and is usually recorded at −18 dBfs digital level for UK broadcast.)

Work tapes should be made in consultation with the sound team, so that tape, format and frame rate are compatible. The edit should be 'cleaned' of any audio that is definitely not wanted, and the audio should be panned hard left and right so that all music and fx are on track 1, leaving track 2 clear for the production sound. TV projects are usually laid down with the first field of the first frame starting at 10:00:00:00. Film projects should be laid down using the reel number as the hour for the start timecode (e.g. the first field of the first frame of reel 1 should start at 01:00:00:00). It is always preferable to use a separate tape for each part or reel. Film projects are more easily handled if split into *double reels* (no more than 20 minutes in length), and each reel should have sync plops at the head and tail. For PAL/film projects, tapes should be striped with 25-frame code, even if the pictures are running at 24 fps (24 fps pull-down).

Viewing tapes are generally played out onto VHS. It is useful to record with burnt-in timecode so that viewing notes can be made with a timecode reference. The relative levels between tracks may need to be adjusted so that the rough mix is not too dynamic and dialogue remains clear. The output level of the mix will need to be dropped considerably, usually via a small outboard mixer supplied with the editing system. This is due to the mismatch of professional and domestic level equipment: the NLE will have balanced I/O referenced at +4 dB, whereas the VHS decks will have unbalanced I/O operating at −10 dB. The loudest parts of the mix should be checked for possible distortion on the VHS deck prior to recording.

EDL

The NLE can output a video and audio EDL, which is of use when the project is moving to an on-line suite, where picture and sound will be auto-conformed from the source tapes. When the project is moving to an audio facility, it is preferable to copy the sequence, delete the video tracks and make an audio-only EDL. Any clips that have been digitized without timecode or roll ID will fail to conform – these should be OMF'd. (It is useful to keep accurate notes about the source of non-timecode material in case it has to be manually recorded into the DAW at a later date.)

To make an EDL, open the EDL window in the NLE. Avid systems use a program called EDL Manager.

Select the chosen *format* for the EDL from the format menu – this must be compatible with the system that will be used for the conform, as most systems work with only a limited number of formats (e.g. CMX 3600, Sony 9000, etc.). The *sort mode* setting controls the order in which events appear

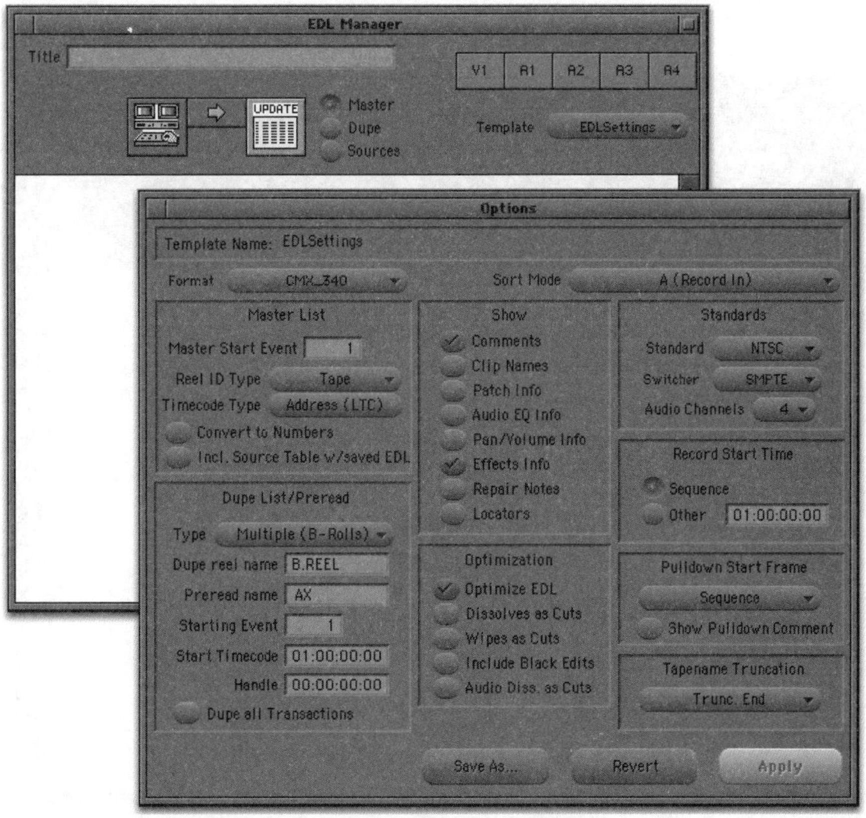

Figure 9.12 EDL Manager settings (courtesy of Avid).

in the list. In A mode the events are listed in the same order as they appear in the sequence. This is not usually the most efficient way to conform a project (unless all the material came from a single source tape), but it is the easiest mode to understand. B mode rearranges the list, grouping all the shots from the same tape together *and* in the order that they appear in the sequence. C mode also groups shots from the same tape together, *and* sorts them in source timecode order. This mode minimizes shuttling back and forth on the source tapes. The *optimization* setting basically simplifies the list. It is best to remove all audio fades and dissolves, replacing them with straight cuts – fades cannot be carried across on the conform. A *handle length* should be set: this adds a number of invaluable cutting frames to the head and tail of each clip. The *standards* setting should be set to PAL or NTSC, whichever is appropriate. Most EDL formats can conform up to a maximum of four audio tracks at a time, so the *audio channel* setting should be set to 4. Where more than four tracks of audio have been laid, separate EDLs for tracks 1–4 and 5–8 should be generated. The *Record Start Time* and *Master Start Event* set the timecode start and first event in the EDL. The Dupe List panel enables the EDL to control a second source machine in the conform: this will be necessary to create transition effects such as a dissolve. Finally, the *Show* panel enables the editor to add information about events. For sound work, the most useful of these is 'clip names': if this is set, slate and take numbers will appear on the EDL. This information can then be used by the sound editor to locate material from the original tapes. Other options in this box, such as Audio Eq and Pan/Volume

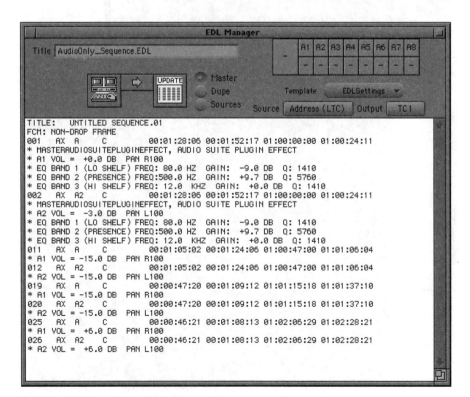

Figure 9.13 An EDL (courtesy of Avid).

Info, will add information about audio settings in the sequence, but will *not* reproduce those settings in the conform.

When all the options have been set correctly, click on the Apply button. This will make an EDL for the sequence, which should now be saved to a floppy disk. EDLs should be delivered using DOS disks, *not* Mac-formatted disks, as many systems will not read these. Filenames must consist of eight characters, followed by .EDL, and it is useful to try and give as much identifying information in the name as possible. For example, R1T1TO4.EDL may be used to describe reel 1, tracks 1–4 (see Chapter 4, 'Linear auto-conforming' section).

Where rushes have been synced by a third party, prior to arriving in the cutting room, the source time-codes in the EDL will not match the timecodes of the original material. Here, the flexfile is used to convert the EDL into a version which refers to the original timecode recorded with the original source material.

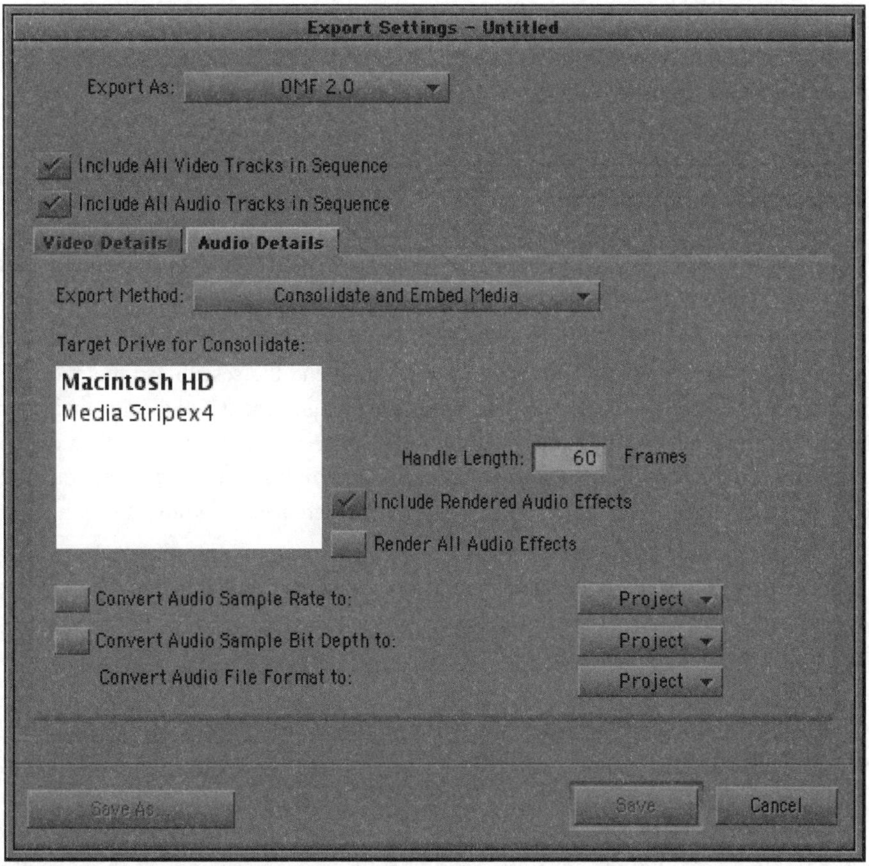

Figure 9.14 OMF Settings window (courtesy of Avid).

OMF

The NLE can output picture and sound, or sound only as an OMF. The sequence should first be copied and the video tracks deleted if the OMF is to be audio only. A handle length should be specified and the sequence then *consolidated*. This means that only media present in the sequence will be copied to the drive, which prevents the OMF from being unnecessarily large. The project can be exported as a *composition-only* OMF (where the file *links* back to the media on the drive) or a *composition with media* (in which case the media is *embedded* in the actual OMF). There are two formats – OMF1 and OMF2. Both handle the audio transfer in the same way, but most editing systems will only read OMF1 or 2, so it is important to check which format is compatible with the destination system.

To make the OMF, first consolidate the sequence and open the OMF Export window. Select OMF 1 or 2. If the video is to be exported, select either linked or embedded video. If the audio is to be embedded, choose WAVE (same as BWAV) or AIFF-C file format – again, this should be chosen with the destination in mind. (If SD2 is offered as an option, it should only be used to export audio to Pro Tools.) If the audio is to be exported as a composition-only OMF, choose the *Link to Audio Media* instead. This will link to the files on the drive in their native format. Once the settings have been confirmed, the OMF should be saved to the location selected, such as a removable SCSI or Firewire drive, DVD or Jazz disc (see Chapter 4).

Spotting the soundtrack

Before sound editors start work on a new project, most productions will organize a 'spotting session' with the director. Depending on the size and complexity of the project, this can be anything from a brief chat about general style to a lengthy and detailed discussion of all aspects of the soundtrack. On a larger project it is usual for the dialogue and fx editor to attend the session, so that they can be fully briefed on a scene-by-scene basis regarding the director's requirements, and offer their own suggestions. One of the sound editors (often the fx editor) will supervise the team from then on, co-ordinating the sound post production team, and acting as the main point of contact with the clients. Detailed notes should be taken during the session, with a timecode reference, regarding the following:

● The general style of the soundtrack (other productions may be alluded to).
● Atmosphere, spot fx and foley requirements.
● Required ADR (revoicing/change in performance).
● Voice-over requirements.
● Areas where music may predominate.
● Crowd requirements.

The initial spotting session is crucial in helping the sound team interpret the director's wishes. Although the spotting notes will be added to, they will constantly be referred to throughout the track-laying period prior to the mix.

Handing over to the sound editors

To summarize, once the picture edit is complete, the cutting room will need to supply the sound editors with the following:

- Guide pictures at the correct frame rate with sync plops, and audio guide track on the specified format.
- An audio EDL saved to floppy disk for each part or reel and all source tapes, if the project needs to be auto-conformed.
- A drive containing the consolidated media for each part or reel if the project is to be OMF'd.
- All associated paperwork, such as marked up scripts, soundsheets, etc.
- Any non-sync source audio, such as archive tapes, music CDs, etc.

10 The digital audio workstation

Hilary Wyatt

An overview

Before computers, the traditional method of placing a sound in sync with picture meant cutting a recording on perforated magnetic stock and joining it with sticky tape onto a roll of perforated spacing the same length as the picture material. When the film and sound rolls were locked and run together, the sound would be played back at the exact frame on which it was placed in alignment with the picture. If the sound was required in the film a second time, the editor would carry out the same mechanical process, assuming that he or she had a second copy of the sound available on magnetic film stock. This was very time-consuming and labour-intensive, and a very *linear* way of working.

Digital audio editing

The digital audio workstation (DAW) enables sound editing to be carried out in a *non-linear* way. A DAW is essentially a computer-controlled system that can record, edit, process and play back audio in sync with picture and other external systems. Most DAWs now offer fully automated on-board mixing, enabling a project to be completed within a single computer environment. Source audio is recorded or imported into the DAW, where it is named as an *audio file*, *region* or *cue*, and saved to a hard drive. The cue is played or *scrubbed* along its length, and in and out points marked. The editor then shuttles along the picture to find the exact frame on which the cue should start. The cue is dragged or *spotted* into position in the *playlist*, where further editing of the cue can take place against picture to ensure that it is *in sync*. Fades and crossfades may be added to smooth out the audio cuts within the cue, and the cue may then be processed using a *plug-in* to add effects such as timestretch or reverb. The process is then repeated with the next cue, building up the audio across the available tracks until the *tracklay* is complete. The playlist is a graphical representation of the audio tracks, and is essentially a list which instructs the DAW when to play each cue and in what order. On replay the computer simply refers back to the original cue held on the hard drive, accessing it as and when it is required to do so. If only a small section of the cue is needed in the edit, then the computer will play back just that section of the original cue without destroying or cutting it.

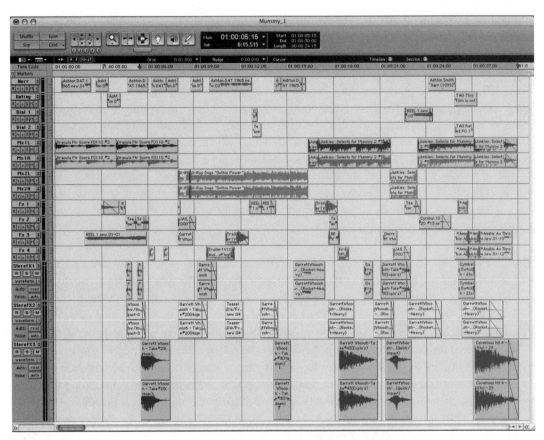

Figure 10.1 A Pro Tools playlist (courtesy of Digidesign and Lauren Waiker).

The process of editing is entirely digital and non-destructive. Edit points can be accessed instantly by clicking on to the selected point within an audio cue, and cues can be inserted, removed, repeated, looped and re-edited at any point during the tracklay. Once the tracklay is complete, it can be exported to a studio or theatre for mixing. Alternatively, the mix may be completed within the system as a mono, stereo or multichannel mix (5.1 or 7.1) and laid back directly to the appropriate format.

System configuration

All DAWs have at their core a computer that houses the audio software. This computer may be custom built, or it may be an 'off-the-shelf' Mac or PC. In either case, the computer must run a specified *operating system* (e.g. Mac OS10, Windows XP, etc.), which will support the DAW software and carry out routine tasks such as updating the graphic display. It must also have at least the minimum amount of *RAM* specified by the DAW software manufacturer. The computer is adapted to carry out audio and video processing tasks using a number of *expansion cards*: the configuration of these can vary from system to system. Most high-end systems will have a *video card*, which will capture picture as a digital file, and replay it with the audio tracks. All systems have a *sound card*, which enables sound to be sent to and from the DAW via analogue and digital inputs and outputs, often referred to as *I/O*. The *DSP card* (digital signal processing) handles all audio processing, such as real-time plug-ins, and mixer operations. The *sync card* enables the system to receive word clock, and to interface with external units via SMPTE timecode, Sony nine-pin remote and other industry standards. The system cards may be housed in an *expansion chassis* or built into the *mainframe* of a custom-built system. Alternatively, some systems use software rather than cards to carry out certain tasks – in this case, some DSP processing is done by the host computer and its CPU, which minimizes the amount of hardware needed.

The term DAW covers a multitude of systems ranging from a simple stand-alone card-based plug-in for a desktop PC or Mac, to a high-end system that is designed to be a 'studio in a box'. However, most DAWs fall into one of two categories:

1. *Host PC- or Mac-based editing systems.* There are an increasing number of systems available that are designed to run on an existing Mac or PC 'host' computer. The user interface is usually a mouse and keyboard, and the graphics are displayed across one or two standard computer monitors. The graphic display varies from system to system, but all rely on drag-down and pop-up menus to access editing functions. These systems tend to be relatively inexpensive both in terms of the DAW software and the computer, usually a standard model adapted for post production with additional software/hardware modules. Software plug-ins are simply Mac or PC compatible; they are generally inexpensive and there is usually a wide choice of software available for most processing functions. This type of system offers high flexibility in terms of hardware and software upgrades. For example, more DSP or extra I/O modules may be added, and the system will benefit from improvements in software made by a variety of manufacturers. For this reason, this type of system configuration is known as *open system architecture*. The downside is that each system is essentially an assembly of components made by a number of manufacturers, and this may sometimes have implications for the smooth running of the system, as well as the quality of after-sales backup, but some of the most powerful DAWs on the market are now produced in this way.

2. *Custom-built systems.* These systems consist of dedicated hardware running proprietary software, all of which is designed and built by a single manufacturer. The advantage of this kind of system is that it has been completely designed to carry out a specific set of operations with a dedicated control surface whose buttons and jog wheels perform specific tasks. The software is often easy to navigate, and typically consists of a record page, an editing page and a cue directory. The CPU is housed in the *mainframe*, which also houses all I/O, DSP cards and associated hardware. These systems tend to be more stable than open systems, but their design is inherently less flexible. These systems are entirely dependent on the research and development of a single company to progress the system further, and implement software upgrades. At some stage these improvements may involve a major hardware/software upgrade that may be prohibitively expensive or indeed not possible within the current confines of the mainframe. For example, the I/O of a system may be physically fixed to 16 or 24 tracks because the mainframe is built that way, and the system may lack the processing power to run 32 or 48 tracks in its current form. The choice of third-party plug-ins may be quite limited and more expensive than the Mac or PC version of the same software, and the systems themselves

are priced towards the higher end of the market. Buyers of custom-built systems tend to deal directly with the manufacturer, which can result in a good after-sales relationship and the possibility of the user having an input into any future software developments. This type of configuration is known as *closed system architecture*.

In any DAW the amount of processing power a system has at its disposal is directly related to the speed of the CPU. This will affect the number of plug-ins that can be run at any one time, and the amount of I/O that the system can run. The number of tracks a system can play at any one time is determined by the amount of memory or RAM available. It is also determined by the speed of the hard drives on which the media is stored, and the speed of the *interface* that connects the drives together, and through which the host computer retrieves the audio data.

Hard drives

Rather than store the media on the computer's own internal drive, modern DAWs can store audio cues on a series of *external* or *removable* drives. A drive may be free standing or designed to sit in a custom-built chassis, and is connected to the computer via an *interface*. Pressing the 'play' button prompts the system to locate and retrieve each cue as it appears in the playlist, and send it via the interface to the RAM, where it is played out by the CPU. The speed at which this happens (measured in milliseconds) is known as *access time*, and the amount of data which can be retrieved at any one time (measured in megabits per second) is known as *throughput*.

There are three common types of hard drive used in post production:

1. *SCSI*. This type of drive has been around for a long time and as a result a number of standards exist which operate at very different speeds. However, SCSI drives still generally have faster access times and higher throughputs than IDE drives, albeit at a higher cost. As a result, SCSI is the most common type of drive used with professional DAWs. Formats in current use range from Wide SCSI (with a throughput of 20 MB/second) to Ultra3 SCSI (with a throughput of 160 MB/second). Drives can be *daisy chained* up a maximum of 16 per system and the *bus*-based SCSI interface can also control other SCSI devices such as a CD-R/W. Drive sizes are typically between 18 and 36 gigabytes.
2. *IDE*. These drives have a slower access time and throughput compared to SCSI drives, with data transfer rates of up to 133 MB/second. A second type of IDE is the ATA drive, which has a maximum throughput of 150 MB, although the newer Serial ATA drives are even faster. IDE drives are installed in most computers, and are Mac and PC compatible. Although cheaper than SCSI, IDE drives are limited to a maximum of four per system, which may not be sufficient in some circumstances. They do, however, tend to be larger than SCSI drives.
3. *Firewire*. Firewire is a high-speed I/O interface with a data transfer rate as high as 800 MB/second (Firewire800). The drive itself is actually an IDE drive built into a Firewire enclosure and, whilst the Firewire interface itself is very fast, the data throughput of the actual drives is a maximum of 40 MB/second. These drives are used on Mac and PC systems and offer low-cost, high-capacity storage. (Drive sizes may typically be 150–250 gigabytes, but can be up to 1 terabyte.) It is possible

to run an extremely high number of drives per system (up to 63). An added advantage is that, unlike SCSI devices, these drives can be plugged up or *hot plugged* whilst the DAW is operational. It is important to note that only *full duplex* Firewire drives are suitable for audio post production. (Full duplex drives enable data to travel in both directions simultaneously – some Firewire drives are only *half duplex*.)

Drive sizes have rapidly expanded to meet the need for increased storage and are continuing to do so. However, where a project is stored on a single very large drive, all the data for that project has to reach the host system via a single interface. A bottleneck will be created each time the amount of data required by the DAW exceeds the speed of the drive, or the capacity of the interface. This may cause the system to slow down, and some tracks may play out incorrectly. Thus, there is a limit to the number of tracks a DAW can play at any one time – this is known as the *track count*. The track count is also affected by bit depth/sampling rates. For example, a system with a maximum track count of 48 tracks at 16-bit/44.1 kHz may only support 12 tracks when working at 24-bit/48 kHz. In the near future, it is likely that the working sample rate may become as high as 24-bit/96 kHz, with ever higher track counts needed to accommodate multichannel audio.

Drive configurations

Where very high sample rates and track counts are required, system performance can be enhanced by saving data across an *array* of drives. This spreads the task of data retrieval amongst all the devices in the array, whose combined throughput will exceed that of a single drive.

Chaining drives

A series of drives can be hung off a single interface using a technique known as *daisy chaining*. Up to 16 SCSI devices can be configured in this way, although certain rules need to be followed. Each device needs to be given a unique SCSI ID number – SCSI ID 0 usually refers to the internal hard drive, and external devices are numbered from ID 1 to 15. The disadvantage of chaining devices together is that

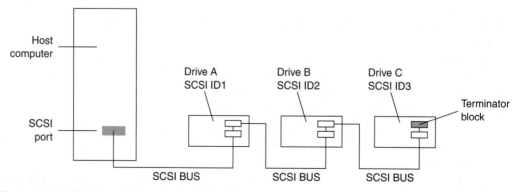

Figure 10.2 SCSI drives 'daisy chained' together.

a device with a low throughput will slow the bus speed of the entire chain down to its own speed. The last device in the chain needs to be *terminated* to prevent the signal being reflected back up the chain. Firewire does not need unique IDs or termination. Up to 16 devices can be chained together, and up to 63 devices can be accessed by creating a *hierarchy*, or via the use of a Firewire *hub*.

Whilst an array of drives will offer some improvement in system performance, it will not *optimize* drive throughput or available storage space, and data may still be transferred via a single interface. Furthermore, drives that are directly attached to a system – known as *locally attached storage* – can only be accessed by that system. This is a fundamental disadvantage where a number of editors are required to work with the same media, which must be copied or saved to another set of drives.

RAID

The alternative to a JBOD array (Just a Bunch Of Disks) is to use a system which applies a level of data management to the storage and retrieval of media. Multiple drives may be striped into a RAID (Redundant Array of Independent Disks) configuration. Here, the operating system 'sees' a number of drives as a single storage device, and data is saved across all drives in an 'intelligent' way that distributes the work of data location and retrieval *equally* between storage devices. Data throughput can be maximized by the use of multiple I/O interfaces managed by a RAID controller. For example, one Serial ATA controller may manage four 250-gigabyte drives (1 terabyte) at a data rate of 150 MB/second.

The drives themselves are partitioned into small units or *striped*, and the stripes are *interleaved* so that when a file is saved it occupies a number of stripes from each drive. This means that, when the file is requested by the operating system, the access time is shorter because each drive has only to retrieve part of the required data.

Most RAID systems offer some level of *system redundancy* by writing files to more than one drive or *mirroring* data. This means that if one drive fails, the data can be transparently recovered from another drive without any visible interruption.

RAID systems can also support one or more *file servers*, which enables files to be shared between a number of systems linked to a network.

Networking and filesharing

The primary advantage of networking is that it allows files to be accessed by multiple users and removes the need to copy media or physically move drives around from system to system. There are various levels of network – from a small server accessed by two or three DAWs to a full-blown Storage Area Network, which can handle the entire acquisition, post production and transmission of media.

Local Area Networks (LANs)

Most DAW manufacturers offer a proprietary networking system that enables a number of systems spread over a relatively small area to access files held on a storage device such as a RAID. For audio

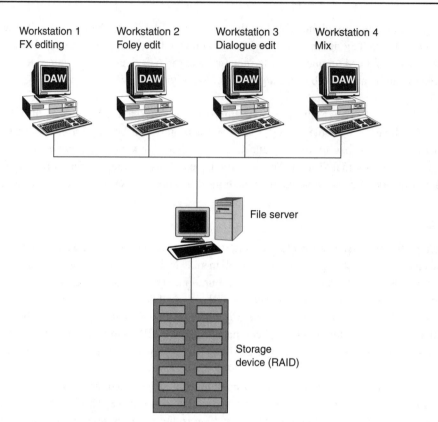

Figure 10.3 Sound editing based around a Local Area Network.

applications, the RAID may typically contain all the effects libraries available to a facility. This is connected to a server, which in turn is connected to a hub that routes data via a *Local Area Network* to each of the attached workstations.

In practice, each workstation sees the RAID as a *network attached storage* device, from which sound files can be copied down to the DAW's locally attached drive. Most LAN-based systems use standard Ethernet technology – either Fast Ethernet (10 MB/second) or Gigabit Ethernet – although some of the available bandwidth is used for system management. As all the workstations must access data through a common shared server, the editing operations carried out on one attached workstation can impact on other network users, and at times where data movement is particularly heavy, a bottleneck may be created which causes the entire system to slow down.

Storage Area Networks (SANs)

Storage Area Networks can transfer multiple channels of video and audio data at very high speeds between RAID arrays and a very high number of simultaneous users. Many SAN systems use Fibre Channel – a fibre-optic bidirectional interface that can transfer data at a rate faster than either Ethernet or SCSI. Fibre Channel can transfer data at up to 2 gigabits/second between devices that may be up to

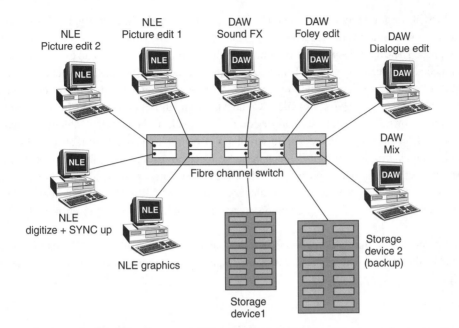

Figure 10.4 Post production based around a Storage Area Network.

6 miles apart over a dedicated network (a 10 Gbps version is in development). The RAID array(s) and SAN hardware are connected directly to a hard drive controller card installed in each attached workstation. Unlike the LAN, the SAN eliminates the need for a shared server and numbers of locally attached drives. Each editing system sees the SAN attached storage in exactly the same way that it would see media saved to a locally attached drive, and all editing is carried out using the shared files directly from the centralized storage system without the need to make a copy first. Because each user has direct access to the RAID, the operations of one editor have no effect on the operational efficiency of the network for other users, and the system is centrally managed by a Fibre Channel switch. All data is mirrored to a second storage device connected directly to the SAN, and all backing up and restoring is carried out invisibly.

In practice, SANs can manage up to several terabytes of media, which makes this kind of network capable of storing all the video and audio files for a large production (such as a feature film), or many smaller projects that can be digitized, edited, tracklaid, dubbed, backed up and even broadcast without actually physically leaving the system. The Avid 'Unity' system, for example, is designed to give up to 60 'clients' real-time access to between 14 and 18 terabytes of media. SAN attached storage can be used in conjunction with Local Area Networks and existing local storage devices.

Working with picture

A DAW designed for film and TV work must support timecode and Sony nine-pin protocol to enable it to lock frame accurately to picture in all playback modes (low-budget DAWs will not support either).

The DAW will use longitudinal timecode (LTC) in playback mode, but for jogging back and forth it refers to the vertical interval timecode (VITC) to maintain sync and park accurately on the correct frame when stopped. The timecode and nine-pin interface allow the DAW to control an external deck or slave to the timecode output of the deck. This arrangement is adequate for tracklaying but does not allow simultaneous scrubbing of audio and video tracks. A better arrangement is to digitize the picture track into the system either as a *QuickTime* file (PC and Mac compatible) or as an *AVI* file (compatible with Windows platforms). The picture will appear on a video track in the tracklay, although it may not be edited. As well as enabling frame-accurate scrubbing of picture and sound, all tracks can be randomly accessed at the click of a mouse. It is much easier to locate an exact sync point using on-line pictures and the editor can check sync on a frame-by-frame basis. A third option is to use a piece of software called *Virtual VTR*, which enables a host computer to emulate a VTR in slave/master mode. This device offers the advantages of on-line video without using up valuable processing power in the DAW.

It is important to note that, although OMFs may contain video and audio, DAWs cannot import the video part of the file and will ignore it. This situation may change once AAF is fully implemented (see Chapter 4).

System requirements and interconnectivity

A DAW is almost never used as a stand-alone system in a professional tracklaying or mixing environment. For this reason, professional systems must offer a high degree of interconnectivity to other systems on a number of levels. Low-end DAWs, on the other hand, are more likely to be used as

Figure 10.5 Virtual VTR (courtesy of Gallery Software).

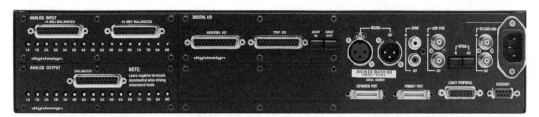

Figure 10.6 Multiple I/O options on Digidesign's 192 I/O interface (courtesy of Digidesign).

stand-alone systems and as a result do not offer the same level of integration. For example, a non-professional DAW may offer four analogue I/Os, S/PDIF connections and word clock only, whereas a high-end system will offer the following options:

● *Audio I/O*. The number of tracks (or *voices*) a DAW can play simultaneously is unrelated to the number of actual input/output channels available. I/O may be housed in a *breakout box* or installed via an *I/O card* slotted into the computer. In the case of custom-built DAWs, the I/O will be situated on the rear panel of the mainframe. All systems will often have analogue I/O, usually two, four or eight mono channels connected via balanced XLR or ¼-inch jack. There are two types of digital I/O – stereo and multichannel.

Most professional DAWs will be equipped with multiple digital connections. There are a number of output types:

  – *AES/EBU* uses either a D-type multipin connector or standard balanced XLR connections, and will receive or output a stereo signal down a single wire at a sampling rate of up to 24-bit/96 kHz or 24-bit/192 kHz using dual wires (also known as *AES 3*). Carries word clock within the audio signal.
  – *ADAT* uses a single fibre-optical connector to receive or output eight tracks at a sampling rate of up to 24-bit/48 kHz (also known as *Lightpipe*). Carries word clock within the audio signal.
  – *MADI* (Multichannel Audio Digital Interface) supports 56 mono or 28 stereo digital audio channels in one direction at either 44.1 or 48 kHz using a single standard BNC cable. Higher resolutions can be supported but this results in a lower track count (e.g. 26 mono channels at 96 kHz).
  – *S/PDIF* (Sony/Phillips Digital Interface Format) uses unbalanced RCA jacks to receive or output a stereo signal at a sampling rate of up to 24-bit/96 kHz. Considered a semi-professional/consumer version of AES/EBU. Carries word clock within the audio signal.
  – *T/DIF* (Tascam Digital Interface Format) uses a single unbalanced DB25 connector to carry eight bidirectional channels of data at a sampling rate of up to 24-bit/48 kHz. This proprietary format was implemented by Tascam for use with their DA88 recorders (see below).

● *Frame rates*. A professional DAW should offer the option to switch between the major broadcast standards and work at the approproate frame rate, including NTSC 29.97 fps non-drop, 29.97 fps drop frame, 30 fps and PAL 25 fps, as well as 24 fps film speed.
● *Sample rates and bit depth*. Until recently, the two industry standards for audio resolution have been 16-bit/44.1 kHz and 16-bit/48 kHz, both of which reflect the maximum data resolution of

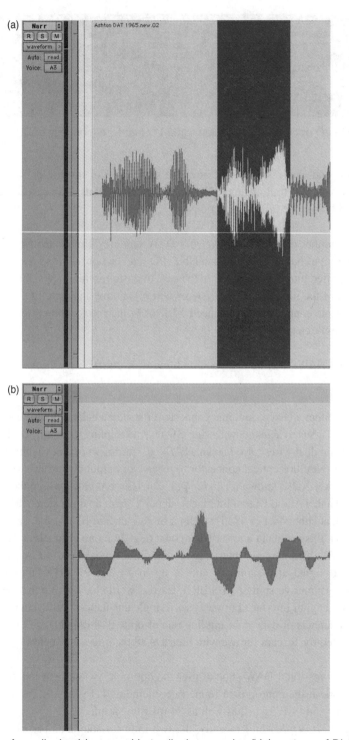

Figure 10.7 Waveform display (a) zoomed in to display samples (b) (courtesy of Digidesign).

audio CD and DAT. However, many systems now operate at 24-bit/48 kHz. The increasing use of file-based field recorders means that location audio can now be recorded at 24-bit/96 kHz, with some recorders able to support a sample rate of 192 kHz. In order to maintain the quality of the original recording, DAWs must be able to work at these higher speeds and some newer systems already do.

● *Sync protocols.* The major protocols required to achieve sync lock in a professional studio environment are LTC, VITC, Sony nine-pin, MIDI and biphase for film work. All digital devices need to be *clocked* from a single source, so that their internal clocks are locked sample accurately. This is known as *word clock*, and some I/O formats such as AES/EBU and ADAT can carry sync information embedded within the audio data. However, in most professional environments a dedicated sync generator will distribute word clock to *all* devices via a BNC connector. Word clock is also known as video ref, house sync or blackburst.

● *File exchange.* File format options should include BWAV, AIFF and may also include SD2. As file transfers are becoming more common, file type compatibility is crucial. Older systems that use proprietary file formats may only read these files after they have been opened in a conversion utility (see Chapter 4). Most systems support OMF, which is frequently used to exchange projects from DAW to DAW, as well as from picture editing systems to DAW. Although newer, AES 31 has been implemented on most systems and is used to exchange audio projects between DAWs. AAF has not yet been implemented at the time of writing but is planned for the near future.

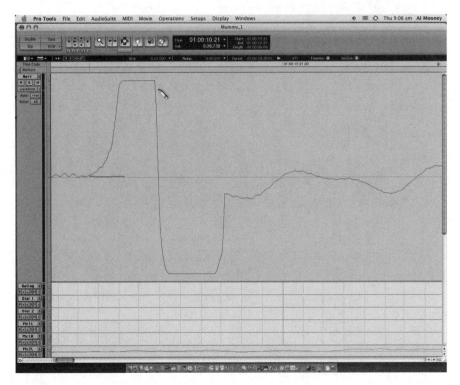

Figure 10.8 Pencil tool redrawing waveform (courtesy of Digidesign).

● *Auto-conform.* Only a handful of high-end systems have the ability to auto-conform from an EDL. Whilst this is still used as a method of recording production sound into a DAW, its significance is decreasing as file transfers become more reliable.

Audio editing tools

Audio software varies so much in design that it is difficult to generalize about the audio tools that may be available in any system. Some of the standard tools include:

● *Directory system.* Some workstations have a well-structured directory system for the storage of audio cues and some do not. In this case, a system should have a fast search engine to locate a cue amongst possibly thousands of audio files. Otherwise, it may be possible to use third-party software such as 'Filemaker Pro'.
● *Audio scrubbing.* This facility enables the editor to jog or shuttle over one or more audio tracks. The more controlled and smoother the movement, the easier it is to locate a precise edit point. Some systems offer scrubbing that slows in speed as a track is magnified or zoomed in.

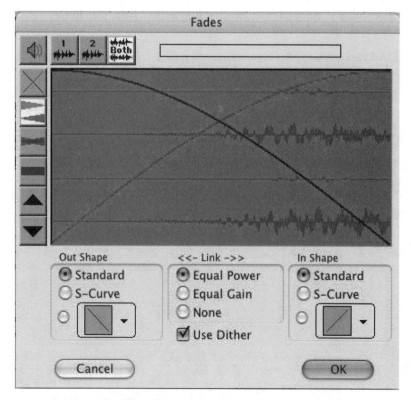

Figure 10.9 Crossfade menu (courtesy of Digidesign).

- *Waveform editing*. In addition to scrubbing, most newer systems offer sample-accurate waveform editing. An edit point may be located by scrubbing and the track *zoomed in* to show the precise detail of the waveform. Where a join is made between two cues, the ideal edit point is where both cues pass through zero on the waveform – any visual step between the outgoing and incoming waveform will be heard as a click. Most systems will allow track and waveform resizing from a *universal view* of the entire playlist down to audio displayed as samples (see Figure 10.7).
- *Pencil tool*. Some systems will allow the user to redraw waveforms (see Figure 10.8). This is extremely useful when removing pops and clicks from production sound. Rather than editing the pop out, it can be located by scrubbing and the track zoomed in to display samples. The pop will show as a visible spike and can be redrawn to match the adjacent areas.
- *Fade/crossfade*. Most systems offer a number of fade/crossfade shape presets, where the length of the fade/crossfade is set by the user (see Figure 10.9). Most systems also offer the facility to unlink both sides of a crossfade so that each can be set independently if required. Some systems create fades in real time, whereas others render fades as new audio cues. A useful feature of many professional workstations is the option to set a very small default crossfade on every cut (e.g. 10 milliseconds). This avoids the possibility of digital clicks on straight cuts.
- *Undo/redo*. All systems will offer a number of levels of undo/redo. This allows the user to try out an edit and revert to the previous version if it doesn't work.
- *Trim tool*. This can be used to lengthen or shorten a cue in the playlist by frame or sub-frame increments.
- *Nudge*. A cue can be moved or *nudged* within a track by user-defined increments of a frame, such as ¼ frame or 10 frames.
- *Autosave*. This feature enables the system to perform background saves at user-defined intervals, e.g. 10 minutes. In the event of a crash, the project will be recoverable by loading the last autosave file. It is good practice, however, to get into the habit of saving regularly, either updating the last edit with all subsequent changes or saving different versions of the edit.
- *Plug-ins*. Most systems come equipped with some standard plug-ins, such as gain, time compression/expansion, eq, reverse and pitch change. Most plug-ins used during editing are *non-real time* and are used to process individual cues. In each case, a new audio file is created or *rendered* – this means that the effect does not use any further processing power on playback.

Mixing tools

Most workstations include a *virtual mixer*, which will closely replicate many of the functions and controls of its hardware equivalent. As the track count of a system can be much higher than the number of output channels, some degree of mixing needs to take place within the system.

- *Routing*. The simplest mixer is one which routes a number of channels into a stereo master output fader. However, most systems will support a number of tracks that can be routed as *groups* or *submixes* into multiple outputs. Some DAWs are designed to accept eight or 12 channels through each input, track, effects send/return, group and output. This means that, in addition to stereo, such systems can support a number of surround formats, such as 5.1 and 7.1. The I/O page of the DAW can be used to create routing presets for a variety of premix and mix formats.

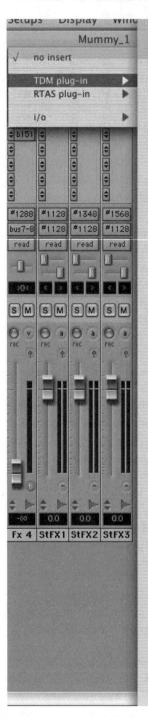

Figure 10.10 Plug-in menu from Pro Tools (courtesy of Digidesign).

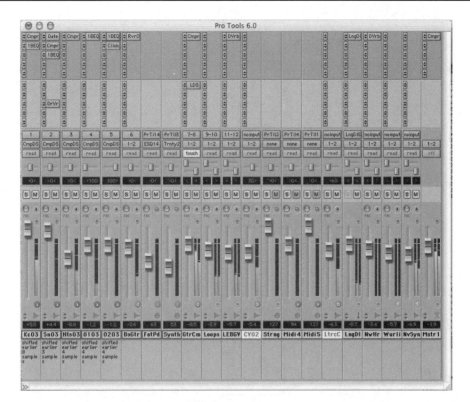

Figure 10.11 An on-board mixer (courtesy of Digidesign).

- *The channel strip*. The channel strip of most virtual mixers is laid out to emulate a hardware strip, with all settings controlled by the mouse. In addition to the channel fader, typical features will include a number of effects inserts, auxiliary sends, stereo/surround panning, gain, solo, mute and automation read/write.
- *Automation*. In *write mode* the system records all real-time mix moves. In *read mode*, the last automation settings are recreated exactly but cannot be altered. Automation can memorize fader movement, panning, real-time plug-in settings and eq. As only one track can be worked on at any one time, automation is a crucial feature of virtual mixers.
- *Eq*. Unlike a hardware desk, eq settings are only called up when required to minimize processing. Typically, systems have a pop-up window containing four band parametric eq settings, as well as a selection of eq and filter presets. All eq settings will be saved in the automation and can be used elsewhere in the project when the same audio characteristics reoccur.
- *Punch-in record*. This feature allows the user to drop in and out of record on the selected record enabled track. To avoid digital clicks, a very short default crossfade can be set at each drop in.
- *Real-time plug-ins*. There are essentially two types of plug in used in mixing – *processors* (such as expanders, limiters, compressors, etc.) and *effects* (such as delay, reverbs, pitch shift, etc.).

Processors and effects that are specific to a single channel can be applied via an *effects insert*. Where a number of plug-ins are applied to a single channel, the signal passes through each one *pre-fader*,

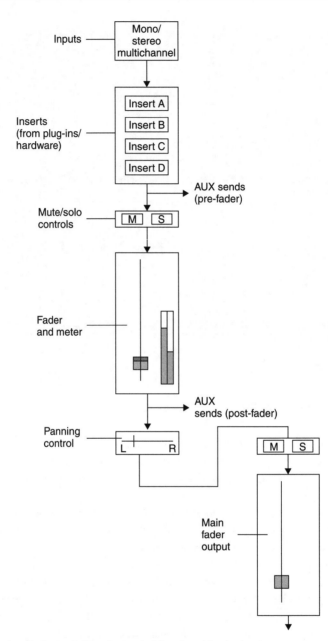

Figure 10.12 Signal path through the channel strip of a virtual mixer.

so the input level of the channel is unaffected by the fader position. Where the same plug-in is used on a number of channels, these can be grouped and the plug-in applied via a group insert. As plug-ins draw on the processing power of the DSP card or the CPU, there is a limit to the number of inserts per channel. Effects (such as reverb) that might be applied to a number of channels, or to the entire mix, can also be processed via the *auxiliary sends*. Here, part of the channel output can be

Figure 10.13 Altiverb: reverb software plug-in (courtesy of Audio Ease).

processed and mixed back with the 'dry' signal via the *auxiliary returns*. This allows the amount of effect to be controlled.

● *Latency*. Every time a signal is processed, small amounts of delay occur. When a signal is repeatedly processed, the delay can build up to several milliseconds. This is known as *latency*. Where the same combination of plug-ins is used across all tracks, the delay will affect all channels equally, thus cancelling itself out. However, where one channel of a multichannel signal is processed, it may end up slightly out of phase. Most plug-ins will specify the delay time in milliseconds, so on a sample-accurate DAW, the processed channel could be manually slipped back into phase. However, many systems come equipped with latency compensation software that enables the user to apply the correct compensation time to the delayed track(s).

In practice, the development of the workstation has bought the jobs of the sound editor and the mixer closer together. Mixing tools are often used in tracklaying and vice versa. For example, gain automation is particularly useful in smoothing out production sound prior to mixing, and delays and reverb may be used by the editor to manipulate sound effects.

Backing up

It is a fact of life that hard drives fail, so it is vital to get into the habit of routinely backing up. Some systems back up media and edit data separately, and others back up the two together. In either case it is important to have two copies on a backup medium appropriate to the system and the project size. Very short projects, such as a commercials mix, can easily fit onto a CD-R. Longer form projects can be backed up to DVD-RAM or DVD-RW, both of which can contain up to 4.7 gigabytes of data. As these are rewritable, either would be ideal for daily backups. Other data storage media used to back up audio include Magneto-Optical (MO) disks, Jazz drives and Exabyte tape. Hard drives have fallen in price so much in recent years that they are now used to archive projects. Whatever the system used, it should be reliable enough to leave running overnight without supervision.

Setting up a tracklaying workspace

Tracklaying may take place in the studio, or in a room which is not necessarily soundproofed. In this case it is important to create the best possible listening environment. With this in mind, all hard drives, computers, mainframes, VTR decks, etc. (in fact anything that emits hum or fan noise) should be installed in a machine room, or failing that, installed in the corridor outside the tracklaying room.

Monitoring

For tracklaying, most editors will choose a pair of high-quality near-field speakers with a separate amplifier: this means that room coloration is minimized. To obtain a good stereo image, the listener and the speakers should be positioned at the points of an equilateral triangle, between 3 and 5 feet apart.

In practice, this means that the DAWs screen(s) is positioned centrally between the speakers. When monitoring in 5.1 surround, the left and right speakers should be positioned as for stereo, and the left surround and right surround placed behind the listener pointing to the listener's head. The centre speaker should be midway between the front speakers, and the sub placed along the left–right axis. The 5.1 set-up will need to be powered by a six-channel amp, or a number of separate amps.

Mixing desk

Most editing set-ups use an outboard mixer to adjust monitoring level throughout the tracklay. If the DAW has an on-board mixer, the output of the master fader can be sent to a stereo input; however, if the DAW does not have an on-board mixer, each track within the tracklay can be assigned a fader so

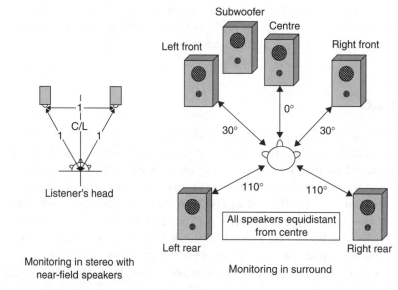

Monitoring in stereo with near-field speakers

Monitoring in surround

Figure 10.14 Monitoring in stereo and surround.

that levels can be adjusted independently. The mixer can also be used to adjust input/output to and from outboard gear, and the mic inputs used to record spot fx and guide voice-over into the workstation.

Outboard equipment

It is helpful to be able to input/output audio directly from any of the most common audio sources, such as CD, Beta, DA88, DAT (preferably timecode DAT) and sound effects server. All decks supporting timecode, particularly DAT, DA88 and Beta, should be connected to the DAW via Sony nine-pin. Other hardware might include a sampler and keyboard, reverb and fx units, although much of this equipment is now being superseded by plug-ins.

All the equipment in the tracklaying room should be led to a *patchbay*. All I/O is connected to a series of ¼-inch jack sockets. Devices are hard-wired vertically in pairs or *normalled*. Once a jack is plugged in, the connection between the two is broken and the other end of the jack lead used to plug in or *over-patch* to another device. This saves a lot of rewiring and is generally kinder to the equipment.

Choosing the right workstation for the job

In most TV companies and independent facilities, the need to move projects and personnel around without raising compatibility issues means that most will install the same system in all areas. The system chosen will be a general all-purpose editor chosen for its ability to interface with client systems, as well as its user friendliness and operational stability.

However, some workstations are more suited to a particular type of editing task than another and, certainly on high-end features, it is not uncommon to see several workstations used during the edit. Prior to the mix, each workstation will be installed in the theatre and slaved to a master controller so that all outputs are locked to a single source, with all systems being controlled remotely by the re-recording mixer. If any editing needs to be done in the mix, the appropriate workstation can be taken off-line for a moment whilst the sound editor makes an adjustment to the edit.

Sound fx editing

Systems must have high track counts to cope with fx laid as stereo pairs and multichannel audio, and must have a sufficient number of outputs to play out a complex tracklay without the need to submix first. Some high-end systems can play hundreds of tracks simultaneously and have as many as 128 output channels available. Sound fx editors like to have the widest choice of plug-ins for sound design work, so a system that can use third-party software 'off the shelf' will be preferable.

Dialogue editing

Track count is not an issue for dialogue editors, as dialogues/ADR can generally be accommodated within 24 or 32 mono tracks. The availability of third-party plug-ins is also not important, although

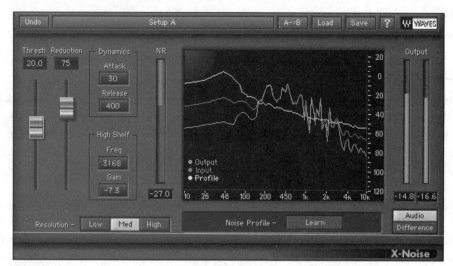

Figure 10.15 Waves X-Noise: a plug-in used to reduce noise and hiss (courtesy of Waves).

some, such as VocALign, have now become part of the 'standard kit' for dialogue editing. This situation is changing, however, because some plug-ins (e.g. *Waves Restoration*) can now emulate the quality of noise reduction only previously achievable within the mixing console. As a result, some editors are using their workstations to do some of the processing tasks that were previously done in the studio.

The most important aspect of a workstation used for dialogue editing is efficient project exchange via OMF, EDL, etc. As it is often not possible to foresee which system a client will use at the picture edit stage, the DAW should ideally be able to cope with all the industry standard project exchange formats.

Music editing

Some workstations were originally designed as music editing systems and have subsequently been adapted for post production work. These systems offer enhanced editing features for music, such as the ability to display the playlist in beats and bars, and the ability to interface with MIDI devices such as samplers and sequencers. As music is often delivered as a surround mix, it is important that the workstation has the ability to import multichannel cues as a single file rather than a number of mono files. This means that the six channels of a 5.1 mix, for example, can be moved around and edited as a group, whilst maintaining sample-accurate sync.

DAWs in the mix

Apart from straightforward playback, the DAW can act as a multitrack recorder or *dubber* in the mix. For this purpose, reliability and system stability are crucial, as is the ability to *drop in* smoothly on record. Some mixing consoles have a DAW built into the control surface, and some DAW manufacturers have

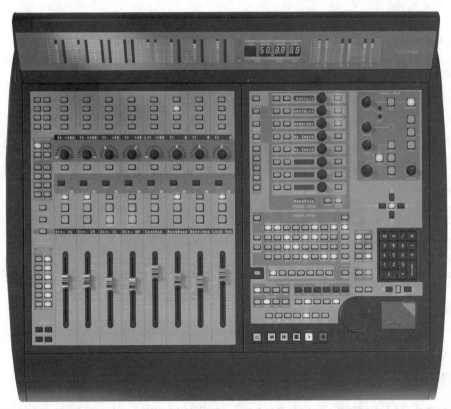

Figure 10.16 Digidesign's 'Procontrol': a hardware mixing interface designed to be used with Pro Tools (courtesy of Digidesign).

developed hardware consoles that work specifically as an extension of the virtual surface (such as Mackies 'Human User Interface' and Pro Tools 'Procontrol').

The main limitation of mixing with a mouse is that only one channel can be adjusted on each pass; however, a physical mixing surface allows a number of channels to be adjusted simultaneously if desired. Additionally, the desk automation can be tied to the audio data, so that when an audio cue moves, the automation moves with it. This is very helpful when reconforming mixes to a new picture cut, as it means the automation is saved and does not have to be remade.

11 Preparing for the mix: editing production sound

Hilary Wyatt

Aims

Whether the project is a simple TV programme that will be tracklaid and mixed by the dubbing mixer in one studio session, or a feature that has a team of editors working on the dialogues, the main aim at this stage is to edit the production sound in such a way that a scene is perceived by the audience to be a continuous piece of audio, much as they would hear it in reality. The editor uses audio editing techniques to make each shot transition inaudible to the ear, masking the fact that each scene is made up of an assembly of discontinuous audio sources. It is an important but essentially invisible art because the audience unconsciously *expects* to hear naturally flowing dialogue. Any poor edits, sync problems, jumps in performance or badly fitted ADR will immediately pull the audience out of the story and interfere with the narrative flow.

This chapter looks at some of the techniques that can be used to make the most of the production sound on any type of project. All the techniques discussed here will be used at some point on a drama or a feature where a team of *dialogue editors* may be engaged in preparing the tracks. To make a scene work more effectively, the production sound may be replaced or creatively augmented with additional ADR lines and voice-over. On high-end TV and film projects, crowd tracks may be recorded and edited to enhance the drama and realism of a scene.

However, good production sound editing skills are equally important on documentary/reality shows, where the original recording will always be used regardless of its quality. Here, the roughness of the audio is understood by the viewer to be part of the 'realism'. This type of project is likely to be edited by the dubbing mixer as the project is mixed. Unlike drama, the possibility of re-recording a scene as ADR is not an option, and much of the mixer's skill will lie in salvaging poor-quality sound that may have been recorded in difficult conditions. At its simplest, this can merely consist of a 'smoothing out' of the production sound, using cross-fades and level alterations between shots. Once a smooth single track has been created for a scene, it is then ready to be filtered and compressed. Alternative takes and wildtracks may be fitted, and voice-over recorded and tracklaid.

The conform

The audio from the NLE will be transferred into the DAW using one of the project exchange formats described in Chapter 4. The final 'locked' audio edit will appear as a series of hard audio cuts (with handles), which follow the picture cuts. The audio level and background noise will change from wide shot to close-up, etc.

As most picture edit systems do not allow sub-frame editing, each audio edit will be on the nearest frameline and will often be audible. Although transfers via OMF allow for the export of some fade and level information, the sound editor should review these edits, as picture editors are often working in less than ideal conditions for audio monitoring.

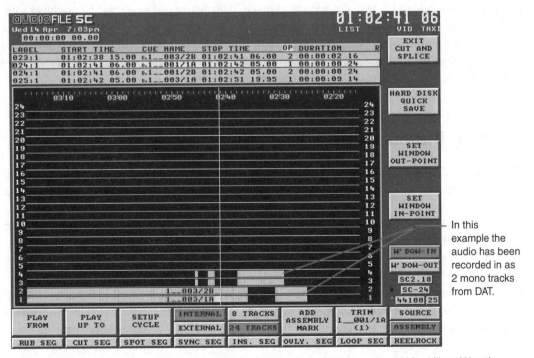

In this example the audio has been recorded in as 2 mono tracks from DAT.

Figure 11.1 A conform loaded into the DAW (courtesy of AMS Neve, adapted by Hilary Wyatt).

Depending on the picture editor's style (and schedule), some alternative takes, wildtracks and sound fx may be already roughly fitted. The majority of the sync sound will have been butt-joined onto one mono or stereo track, and any splitting onto other tracks will probably be to avoid overlaps on the main tracks, or because the editor wishes to keep sound fx/wildtracks and music separate for 'housekeeping' purposes. On projects where a number of mic sources are used, a separate track of audio will appear in the DAW for each source. The conform of a drama or feature may simply have the boom mix on track 1 and radio mic mix on track 2.

Checking sync

Before any work is done on the conform, it must be played against the guide audio track on the picture playouts supplied by the cutting room to establish whether the two tracks phase with each other. Any audio cues that are out of place can be manually *nudged* into sync with the guide. If there appears to be a widespread sync problem, then the problem may lie in the transfer process or the set-up of the DAW.

Unless the project is a fast turnaround studio-based tracklay, the first job of the dialogue editor is to record the entire guide track into the DAW with timecode (and sync plop if present) and place it on a spare audio track in the conform. This will be the sync reference throughout the edit, eliminating the wait for external machines (if used) to lock up and enabling accurate scrubbing of the track to locate sync points. This guide track will also be used to fit the ADR. Just to make sure that all machines were 'locked' during recording, it is worth playing the head and tail of the guide track against the picture guide track to ensure that they phase together. Any sync errors at this stage will be time-consuming to fix later. Projects that have been auto-conformed will only be sync accurate to the nearest frame. Audio imported as a file transfer (e.g. OMF) should be sample accurate.

Starting the dialogue edit

The dialogue editor will need to be able to locate alternative takes, wildtracks and find 'fill' to cover ADR, and so must have access to all the source material, which may arrive on a set of DATs or a drive. It is useful to load all available wildtracks before starting to edit. The editor will need a corresponding set of sound report sheets and a copy of the 'marked up' script. This indicates how a scene has been shot and is marked up with the corresponding slate numbers. Finally, a printout of the audio EDL is useful, as this will list the actual takes used in the edit.

There are no hard and fast rules about how a scene should be edited, as each will throw up a different set of problems. A domestic interior scene cutting between two actors should be relatively simple to smooth out, as the two angles will be fairly similar. However, an exterior scene shot on a street with the same two actors may result in one angle having a heavier traffic background than the other. The traffic will then step up and down on the picture cuts between the two. A general approach might be as follows:

- Analyse the scene to determine what is the predominant, or noisiest, background – this will be extended and smoothed to create a continuous background presence. Split off any audio cues that differ in background (or recording quality) onto a separate track.
- Separate off any unwanted or unusable audio onto a 'strip track' – often a pair specially assigned for this purpose.
- Split off any lines marked down for ADR onto a separate track reserved for this purpose, so that they are available if the ADR is not satisfactory.
- Split off any sound fx that are clear of dialogue onto a separate track reserved for this purpose. The fx are then easily available to the mixer when the 'M&E' (Music and Effects) track is made and available for the main mix itself.
- The gaps now created need to be filled with matching background noise to create a continuous track. Trim out the handles of the remaining cues to reveal any available 'fill'. If there is not sufficient, it will probably be necessary to load the uncut take into the edit page and find areas of matching fill – often the quiet moment before 'action' is a good source. Check whether a *room tone* was recorded for the scene – if it was recorded from the same angle as the takes used in the scene, then this will be a useful source of fill. The last resort is to record in all available takes from the same angle in the hope that enough fill will be found. This can then be edited into the track and smoothed with crossfades.
- The lines that were split onto a separate track in stage 1 can also be filled and smoothed in this way. An overlap of fill should be created between the two resulting tracks, so that the mixer has an opportunity to blend the two backgrounds in the least obtrusive way. However, if there are only few isolated lines which stand out from the rest of the scene, it may be best not to fill but to minimize the background change. Trim hard against the incoming dialogue and create a long fade at the end of the line – the incoming noise will be masked by the (louder) dialogue, and the fade allows the mixer to gate the line without clipping the dialogue.
- Cut out clicks, bumps or unwanted fx. Words, syllables and part syllables (phonemes) can be replaced from other takes – ensure that the fix does not change the tone of the performance, and that the new audio matches the original line in perspective, pitch and level.
- Where there is a cut from wide shot to a closer angle, replace the wide sound with takes from the close-up, using the original to match sync. Not only will the dialogue be cleaner, but using the same perspective sound to bridge the picture cut does not call attention to the shot change. If an actor is consistent from take to take, then it is often quite easy to replace complete lines convincingly.
- Where a wildtrack is available, or a cleaner take found for a noisy line, these should be fitted and laid on a separate track as alternatives.

Figure 11.2 shows how a conformed scene might look after filling and smoothing out. However, in some locations, the background noise may change considerably from take to take, and smoothing out may not be effective. If ADR of even one angle is not an option, then all the editor can do is split the scene across several tracks, and place overlapping ramps of matching fill at the head and tail of each cue, for ease of mixing.

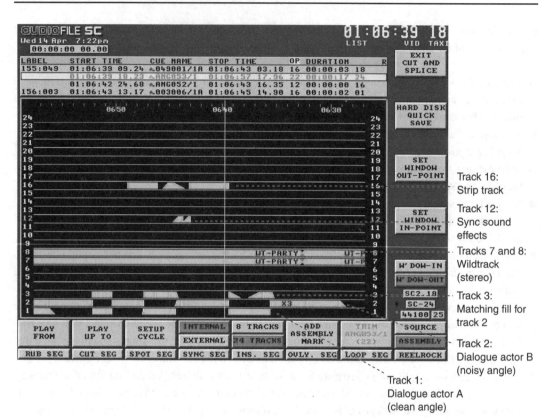

Figure 11.2 A filled and edited scene (courtesy of AMS Neve, adapted by Hilary Wyatt).

Boom or personal mic?

Often a scene will be covered by more than one mic, which does give the dialogue editor some choice. If a scene is covered by a boom and personal mic mix, the editor will need to decide which one is best to use. Often a personal mic is used when the boom swinger cannot get in close enough, resulting in a 'thin' and roomy boom track. However, personal mic tracks can be plagued by clothing noise due to bad positioning, or can sound inappropriately close in perspective because the actor's breaths, movements and even heartbeats may be recorded, as well as their dialogue! Unlike the boom, personal mics do not pick up the acoustic characteristics of the location, which can sometimes be a good thing. However, if the boom track is well recorded, it is much better to use this, rather than having to recreate the acoustic environment from scratch in the mix using artificial reverbs.

Because boom and personal mics have such different characteristics, it is best to avoid switching from one to the other within a scene. Where this is not possible, the two sources should be split onto separate tracks, as they will require a different eq and reverb setting in the mix. In a scene where the boom track is mainly used, matching background fill from the boom should be laid under the personal mic line to help disguise the crossover between the two sources.

Handling twin/multiple-track material

Where the recordist has allocated an audio track for each actor, it will be necessary to clean up each track individually. Start by cutting out noise and crosstalk between each line of dialogue on both tracks. This will result in a scene that is 'checkerboarded' across two tracks. Then fill as necessary, and add fade ins and outs to smooth the transition between tracks.

With the advent of file-based recording formats, track separation is now possible over more than two tracks. Although it will take longer, the same twin-track method can be used to edit audio recorded across a number of tracks.

Handling M/S recordings

M/S recording is a useful technique for recording crowd scenes where a wide stereo effect is desired. However, some sound recordists deliver main dialogues recorded in this way. As the main dialogues will be mixed in mono, it is best to split off the 'side' leg onto a strip track. This contains the stereo left–right part of the signal, and often consists of location noise and *roominess*, neither of which are of much interest to the dialogue editor. Editing the scene can then take place using the 'mid' leg, which is effectively the mono part of the signal. (M/S recording is discussed in Chapter 8.)

Techniques for improving audio edits

Dialogue editing is largely about problem solving, and the more tricks an editor can throw at a scene the better the result will be. Some of the more commonly used ones are:

● Move the edit point off the picture cut to just behind the first frame of the incoming dialogue. The relative loudness of the incoming dialogue will mask the edit because the brain registers the louder

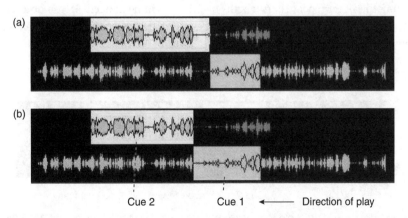

Figure 11.3 Edit detail: (a) original edit position; (b) edit moved just behind incoming dialogue (courtesy of AMS Neve, adapted by Hilary Wyatt).

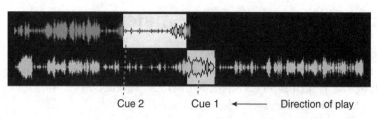

Cue 2   Cue 1   ◄————   Direction of play

Figure 11.4 Edit detail: hiding the edit between two takes within a word (courtesy of AMS Neve, adapted by Hilary Wyatt).

sound at the expense of quieter sounds immediately before and after, such as edits in low-level background noise.

● Mentally breaking down a line of dialogue into a series of syllables, or even phonemes, will reveal its natural edit points, e.g.

I/t i/s par/ti/cu/lar/ly ho/t

● Hard consonants such as p/t/k/b provide the easiest cut points within words. Long vowel sounds, particularly diphthong vowels, are the most difficult sounds in which to make a smooth edit.
● Make your edit at the least obvious point. When joining two dialogue takes together, try to conceal the edit in the middle of a word – it is less likely to be noticed (see Figure 11.4).
● Where a picture cut occurs during dialogue, move the corresponding audio edit off the cut. The viewer is more likely to perceive the dialogue as continuous if the picture edit and audio edit do not happen simultaneously.
● English speakers often produce a glottal stop at the end of a word – most commonly, the final 't' is not articulated. Find a 't' from elsewhere in the scene and edit it into the appropriate place
● Sounds that are created without using the vocal chords, such as 'k', 'f' and 's', can be taken from a *different* actor in the scene, if necessary, and edited in to the required place. These sounds are created with the mouth and do not contain any vocal characteristics, so the change in voice goes undetected.
● Where an unwanted dialogue overlap occurs, try to find a clean start or end sound from other takes to replace the original.
● When editing fill, never repeat short lengths of fill – it will be noticed! If you do need to loop fill, which changes in tone or level over the length, reverse copy the incoming fill so that you are cross-fading between the same tone/level.

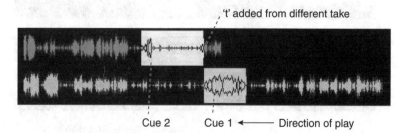

't' added from different take

Cue 2   Cue 1   ◄————   Direction of play

Figure 11.5 Edit detail: adding a cleaner 't' to replace a glottal stop (courtesy of AMS Neve, adapted by Hilary Wyatt).

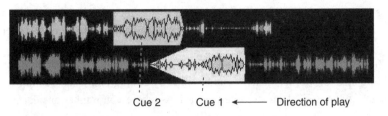

Cue 2   Cue 1 ←———— Direction of play

Figure 11.6 Edit detail: using asymmetric fades to smooth the transition between two takes (courtesy of AMS Neve, adapted by Hilary Wyatt).

- Asymmetric fades are most useful in smoothing background noise between takes. Extending the outgoing background under the incoming dialogue makes use of the 'masking' effect mentioned earlier and creates a smoother transition.
- When removing lines that are to be post synced, leave any sync lip noise/breaths around the line *in the track*. There is always at least one frame of silence between an intake of breath and the start of the line, so cut and fill from this point. The breath will lead up to the ADR line, with the matching fill creating a smooth transition between the two. This will help the ADR to sit in the production sound.
- Where a scene needs extensive cleaning up and refitting, record all the takes in before starting to edit. This may seem tedious, but it does pay off. If there appears to be no hope for a scene and ADR seems the only answer, it is always worth checking *all* the unused takes, because there may just be a more projected performance, the traffic may have quietened down, or the fridge may have been turned off for a later take! This may be fittable and provide, at the very least, an alternative to the ADR in the mix. If the sync is good and the director is happy with the performance, there may be no need to ADR it at all.

Dialogue editing software

There are a number of software plug-ins available, most of which are noise reduction processors. Cedar Audio, for example, sells a software bundle including 'Auto Declick', 'Auto Decrackle' and 'Dehiss': these may be used to help salvage difficult material. It is worth noting that if processing software is used to treat dialogue *prior* to the mix, it is best to retain the original untreated cue in the tracklay. It is often difficult to judge the result in the environment of a prep room, and better results may be achieved in the studio.

ADR spotting

Because of increasingly tight schedules and the need to book actors well in advance for ADR, there is often a need to produce a draft ADR list as quickly as possible once the original audio is conformed. In addition to the ADR lines requested by the director for performance (see 'Spotting the soundtrack' section in Chapter 9), the dialogue editor will review the conform and mark down lines which should be re-recorded for technical reasons. If the production mixer has recorded a scene as twin track, or boom and radio mic, it is important to listen to both legs separately, as they will be quite different in quality.

There are no hard and fast rules for deciding which dialogue should be post synced: it is largely a matter of building up a knowledge of what can be fixed by editing and what techniques can be used in the mix to 'clean up' the track. Here are a few occasions where you might consider it:

- Ratio of background noise to level of dialogue is such that the dialogue is fighting to be heard (e.g. a scene in an airport where the actors cannot project sufficiently over runway noise, or alternatively an intimate scene shot in close-up with very low level dialogue and a high level of camera noise).
- Background noise is inappropriate to the scene (e.g. contemporary skylines in period drama).
- Heavy background change on a particular camera angle – it may be worth post syncing the worst affected lines within a scene, and smoothing out and filling the remaining lines so that the ADR lines will play on top of the sync track.
- Extraneous noise over dialogue from filming equipment/planes/crew/props.
- Sync fx are disproportionately loud in relation to the dialogue (e.g. footsteps or door opening).
- Where the actor is off mic.
- Where the actor on camera is overlapped by an actor delivering off camera, and therefore off mic lines.
- Where the recording quality is not acceptable (e.g. distorted or unacceptably 'roomy', clothes rustle on radio mics).
- Where shots have been filmed to playback, as guide track only, or shot mute.
- Where the sync for a wide shot cannot be replaced with a take from a closer perspective.
- Where the end of a line marked for ADR overlaps the incoming dialogue, then the overlapped line will need to be post synced too.
- Where alternative 'unacceptable language' lines for airline versions are needed.
- Revoicing foreign versions.

If only part of a scene is marked for ADR, it is necessary to decide at what point within the scene should you cross back over to using sync sound. (Where a whole scene or single line is to be post synced, then this issue is much simpler!) If it is necessary to do only part of a speech, for example, then you should post sync the dialogue up to the point you actually need it, *and then continue* as far as the next perspective cut, change in camera angle or natural pause. When mixed, this should help the transition from ADR to incoming sync, as there will be a picture change that justifies a corresponding change in the sound.

Some dialogue marked down for re-recording *will* be successfully fixed during the edit, and these lines can be deleted from the ADR list if it hasn't yet been recorded. Indeed, the mark of a good dialogue editor is his or her ability to repair lines from unused takes in a way which is sensitive to the original performance, without automatically resorting to ADR.

The actual length of each line or 'loop' should be be determined by what the actor might reasonably manage in one take, breaking down the original dialogue in a way that takes into account the natural stops and starts of the original performance.

ADR cue sheets

ADR lines for each actor should be compiled onto an ADR cue sheet.

Each loop is given a reference number (column 1). Column 2 shows the actual programme timecode on which each line starts and which will eventually be the cue for the actor to start speaking. Any breaths or non-verbal noises needed before the line can be put in brackets. Each ADR cue needs to be frame accurate, as any inaccuracy will throw the actors off sync on every take, and frequent recueing of ADR lines in the studio will not help the session go smoothly. The line to be looped is listed in column 3 and should be an accurate transcription of what was actually said on the take, *not* what is written in the script. If breaths and pauses are indicated, this can help the actor reproduce the original rhythm of the line. Column 5 is blank and will be filled with take numbers in the session, as well as a note of the selected takes to be used. On the example shown, column 4 has been used to list the reason why the line is being post synced. If the suggestion came from the director or picture editor, then the abbreviation CR (cutting room) has been used, but if the reason is different (e.g 'bang on mic'), then this acts as a useful reminder during subsequent discussions with the production team.

Once a comprehensive list of ADR has been compiled, it is the director's decision (with some input from the producers) as to how many of the suggested lines will actually be post synced. It is often felt that ADR never recaptures the energy of the scene, and that something is lost from the original performance. This has to be balanced against how much poor-quality audio may detract from that performance.

ADR CUE SHEET

Production	Episode 1
Actor	John B
Character	James

Loop no.	Timecode	Cue	Notes	Take
J1	01 00 34 23	Hi ... is John there?	Off mic	
J2	01 00 56 18	(*Sigh*) ... Well when will he be back?	Off mic	
J3	01 01 40 19	Ok Could you tell him I called?	Low level	
J4	01 01 42 00	(*Laugh*) Yeah.. Ok, thanks..bye!	Creaky floor	
J5	01 01 59 22	I don't know	Cutting Room performance change	
J6	01 02 02 03	running breaths to 01 02 19 11	Radio mic crackle	

Figure 11.7 ADR cue sheet.

In scenes where the recording quality is borderline in terms of acceptability, it is safer to have the edited ADR as well as the edited original production sound in the mix, so that if the latter cannot be salvaged, the ADR is there to fall back on. It is true to say, however, that many TV productions do not have the budget or the schedule to record extensive amounts of ADR, and the dialogue editor will have to enter into a negotiation with the production team about which lines have priority.

ADR spotting software

The process described above is a manual process of scrubbing the sync track to identify the first frame of each loop and typing up the information on a cue sheet. However, there are a number of ADR software packages available which automate this process by following timecode, so that in and out points can be logged by a single keypress. Different versions of the list can also be generated for the actor, director, etc. A system such as the Pro Tools-based 'ADRStudio' then uses the list to manage the session, controlling the streamer, arming the record tracks and naming the cues.

Attending the ADR session

The dialogue editor should liaise with the studio beforehand to specify the sample rate/frame rate of the job, the format on which the ADR will be delivered and any special equipment or microphone requirements. It is often useful to let the studio have the ADR sheets in advance, so that the first set of cues can be put into the streamer before the session starts.

The most important function of the dialogue editor in the session is to assess whether each take is close enough in sync and able to be fitted. The editor should also assist the actor and director in matching the original performance, paying close attention to pitch, projection and pronunciation. The actor should also be encouraged to emulate the energy and movement of the original by physically moving whilst recording the line. (Beware of picking up cloth rustles and footsteps on the recording!) Where an ADR session is used to *change* an actor's performance, the editor should ensure the projection and sync of the new lines are still an appropriate match for the on-camera performance.

The surest way for the editor to check sync during a take is to listen simultaneously to the production sound through a single headphone provided for the purpose. Any difference in rhythm between the ADR line and the original will then be apparent. It is a good idea to ask the ADR recordist to assemble the selected takes onto one track of the DAW, leaving all other takes available on other tracks, as this will speed up the process of fitting later on. ADR recording itself is dealt with Chapter 13.

Editing ADR

If there are only a few lines of ADR to be fitted, it is quicker to place them straight into the main dialogue edit and fit them in situ. If there are a large number of cues to be fitted, it is generally easier to edit the ADR separately to the dialogue tracks, merging the two sets of tracks together once editing is complete.

To fit ADR accurately, it should be fitted to the guide track that was recorded into the DAW at the start of the dialogue edit. It is not possible to fit ADR accurately by eye alone. The exception to this is animation, which is much less critical in terms of lipsync, and where the relationship between the character and the voice is clearly 'created'.

Unless the first take was perfect, there will be a number of takes available to the editor. One or more of these will be the director's selected take: alternatives may be asked for, or a line may be fitted from several takes to get the best fit. Whilst it is important to go with the director's choices, it may be necessary to use a syllable or a word from another take to obtain a better fit or a smoother edit, so it is always useful to do 'one for safety', even if the first take looks great. Here are a few guidelines on fitting ADR:

- It is easier to shorten ADR that is too long than to stretch ADR that is too short.
- Even the best ADR lines need fitting. This means matching to the guide on a word-by-word, syllable by-syllable basis. Fitted ADR is full of edits!
- Start by cutting on the most obvious sync points first (hard consonants such as 'p', 't' and 'k') and match these to the guide. Sounds made by opening the lips from closed (such as 'm' and plosives such as 'd' and 'b') are a dead giveaway when not quite in sync.
- The intervening syllables can then be edited to fit. It is often necessary to 'flex' parts of words by small amounts to get an exact fit, using a timeflex function on the DAW, or an application such as 'VocALign' (see below).
- Constant sounds (such as 'f', 'sh' and 'v') are easy to cut into. Changing sounds (such as the 'aee' of plate) are the hardest points at which to make a good edit.
- Most ADR edits should be straight cuts. Overlaps created by overlong ADR should be dealt with by cutting and/or flexing the line to fit, and *not by crossfading*, which will create an audible doubling up of the actor's voice.
- Tracklay any obvious inhales/exhales/lipsmacks, etc. around the ADR lines either from other takes or from a wildtrack of non-verbal sounds specially recorded by the actor at the end of the ADR session. This will help the ADR lines to appear more like natural speech and less like ADR.
- Where possible, leave some time between actually fitting the ADR and then finally reviewing it – fine-tuning the ADR with fresh eyes can help considerably.

Once fitted, the ADR should be split for scene changes/perspective across the tracks allocated for post sync (see 'Splitting the dialogues for the mix' below) and will be merged into the main dialogue edit prior to the mix.

ADR fitting software

There are a number of 'flexing' applications available that can compress or expand dialogue without altering the original pitch. There are also applications (most notably 'VocALign') that will flex a line of replacement dialogue to match the waveform of the original. Figure 11.8 shows the VocALign window, with the original line in the upper part and the flexed line in the lower part. There are a number of processing options available which will affect how the dialogue is flexed and which should be experimented with for best results. It is important to note that using such software does not lessen the need to concentrate on sync during ADR recording. Indeed, the best results may be obtained from a combined approach, i.e. processing an already fitted ADR line to fine-tune the sync. Such software can also be used to fit alternate sync takes, and to replace one actor's lines with the voice of a different actor.

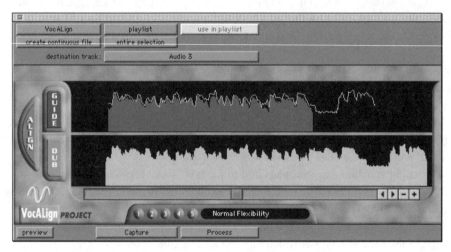

Figure 11.8 VocALign (courtesy of Synchro Arts).

Splitting the dialogues for the mix

Before the tracks can be delivered to the mix, they should be laid out across the available tracks in a way that takes into account *how* they will be mixed. The conventions for this can vary, depending on the type of project:

● Tracks should be laid out so that the mixer can audition outtakes, alternatives, wildtracks and ADR against the main tracks, simply by lifting a fader.
● Tracks should be split for perspective and interior/exterior scene changes, as each picture change will require a new eq and reverb setting.
● Any dialogue that requires a particular effect, such as TV or telephone eq, should be moved to a track assigned for the purpose.
● It may be necessary to split the edited sync tracks by character if vocal characteristics are markedly different. For example, a deep-voiced male who projects strongly will require a different eq setting

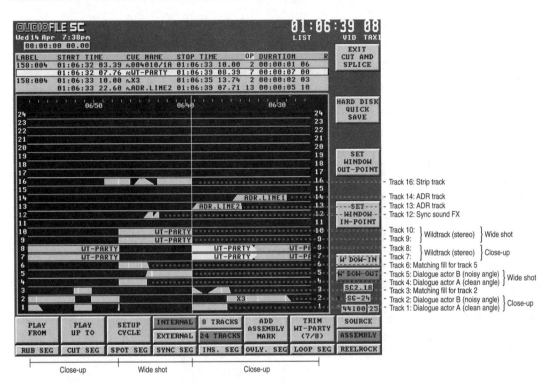

Figure 11.9 A scene split, cut for perspective, and ready to mix.

to a softly spoken woman whose voice is mainly mid-range. This is common practice on feature films.

● If there is a sudden increase in level within a scene, e.g. a scream, it is a good idea to split this off onto another track, cutting hard on the first frame. The mixer then has a visual cue and can anticipate the increase in level with the fader.

Once the dialogue split is complete, the ADR edit can be merged into the tracks. A final check should be made to ensure that the ADR tracks work with the corresponding fill. The edit is now ready to be delivered to the mix.

Crowd spotting

Where principal actors are delivering dialogue in crowd scenes, the extras will be mouthing rather than actually speaking, so that the main dialogue is recorded 'clean'. Although the recordist may go on to shoot a crowd wildtrack to cover the scene, and the fx editor may lay some general crowd atmos, other recordings will probably be required to create crowd backgrounds that are specific to the scene. These are recorded in a post sync session by a group of specialist crowd artists (often known as a loop group). How much crowd is necessary depends on the nature of the production, as well as the production budget. A period film with a number of battle scenes will need a very large loop group over a period of days; a contemporary drama for TV might require six actors for half a day just to cover pub interiors, police stations, etc. Good crowd works almost as a sound effect and should not distract from the main dialogues or compromise intelligibility. Crowd sessions can be used to cover a number of dialogue 'effects', such as:

- Murmur tracks – tracks of non-verbal murmur which are very useful to indicate the *presence* of large numbers of people in a big space (e.g. a church congregation, a crowd at a football match).
- Buzz tracks – tracks of general crowd atmos which can be used as a 'bed' throughout a scene (e.g. party chat).
- Reaction tracks – where a crowd is visibly reacting to onscreen action (e.g. audience cheers, crowds at a horserace, onlookers at the scene of an accident).
- Foreground vocals for specific extras – in small groups or singly, these should be recorded separately to the buzz track so that the eq/panning and level settings can be controlled independently for each group (e.g. two men talking, walk screen left to right).
- Speech which can be treated as a sound effect, such as telephone operators, train announcements, speaking clocks, emergency services, r/t chat, etc.
- Non-verbal dialogue, such as grunts of effort/breaths/sighs, which are missing from the principal actor's performance and which cannot be covered by the actual actor. (This is useful in a situation where the principal actor is only available for a short ADR session.)
- Revoicing minor characters.
- Voicing non-speaking extras who may be expected to interact with principals, such as waiters, servants, etc.

Crowd can be spotted on an ADR cue sheet. The editor should note the incoming timecode, a description of what is required at that point (e.g. pub b/g chat), and lastly the actors required for the loop (e.g. mixed group: all). Once spotting is completed, an estimate should be made as to how much studio time will be required and how many artists will be needed. It is important to specify the breakdown of the group, in terms of age, gender, regional accent, class, etc. If any of the required crowd needs to be scripted in advance of the session, this should also be noted. These requirements are then passed on to the production office, so that a booking can be made with the artists.

Attending the crowd session

The session can be run very much like an ADR session, using a streamer to cue each loop. It is not usual for the director to be present at crowd sessions, and much of the direction will come from the dialogue editor, who should have a clear idea of what is required. In order to run the session quickly and efficiently, one member of the loop group will often act as a 'floor manager', accepting directions from the editor and choosing the most appropriate members of the group to perform each cue. The editor needs to ensure that the crowd is recorded with the correct levels of projection and energy, as well as the appropriate number of voices for the scene. When recording a buzz track, ask the group to vary their projection and leave occasional silences in conversations. This emulates the natural modulations in speech patterns and makes it much easier to blend the crowd with the main dialogues in the final mix. To create the illusion of greater numbers, record two similar buzz tracks, which can then be doubled up to 'thicken' the crowd. (Crowd recording is dealt with in Chapter 13.)

Editing the crowd

Unless the crowd has been recorded to a specific picture reference, it should be tracklaid in a manner similar to the methods used when laying sound fx. This involves being selective about which part of a loop is used and editing it into the best position around the dialogues. Background crowd can be multilayered with foreground detail to create spatial depth within a scene, and some thought should be given to laying up material that may be panned left to right. Distinctive crowd can be usefully laid on the establishing shot, where it will help 'set' the scene, and remain clear of incoming dialogue.

The edited tracks should be split for scene changes/perspective/eq treatment in the same way as the ADR and sync dialogue tracks (see 'Splitting the dialogues for the mix' above). If very little crowd is required for a project, it can be edited and placed on a free track within the main dialogue edit. If the crowd is spread over a number of tracks, however, it may be premixed separately, and for ease of working should not be merged with the dialogue edit.

<table>
<tr><td>12</td></tr>
</table>

12 Preparing for the mix: sound effects editing

Hilary Wyatt

Aims

Whether the project is a documentary that will be tracklaid and mixed in one studio session, or a feature that will have a complex tracklay carried out by a team of editors, the main aim of the sound fx editor is to create an audio landscape which draws the audience into the 'reality' created by the director and out of the environment in which they are actually viewing the film. Sound effects have a number of other functions within the soundtrack:

- Sound effects can create the illusion of reality. Silent archive footage in a war documentary will be perceived differently with the addition of a few sound fx of machine-gun fire and explosions. A studio set will be less likely to be perceived as a set if the 'world outside' is suggested by street noise, children playing, etc., and the interior is suggested by room tones, fridge hums, etc.
- Sound effects can create the illusion of continuity. The fact that a scene is made up of a number of discontinuous shots edited together will be disguised by continuous atmosphere tracks laid over the length of the scene, which the audience will then perceive as a 'whole'.
- Sound effects can create the illusion of spatial depth. Effects that are recorded with the appropriate natural perspective can suggest distance from foreground action – distant gunfire has a very different acoustic to close gunfire.
- Sound effects can create the illusion of space. Effects can be panned left/right or mixed in stereo/surround, creating the illusion of three-dimensional sounds, which is more akin to how sound is heard in reality.
- Offscreen action can be 'told' by sound effects rather than being actually shown. The classic example of this is the offscreen car crash – sometimes a result of budget limitations rather than the desire to avoid being too literal!
- Sound effects can help fix problem areas in the dialogue track. For example, a fridge hum tracklaid over a kitchen scene may help mask an otherwise noticeable lighting hum, as they are both of a similar frequency.

- Any visual stunts, such as breaking glass, need to be tracklaid using the appropriate effect, and the 'prop' effect removed.
- Sound effects can assist in setting mood – for example, rain instead of birdsong on an interior scene.
- Sound effects, together with music, provide the 'dynamic' elements of a mix. Put simply, loud sounds are most dramatically effective when juxtaposed with quiet moments in the soundtrack, and vice versa.

Types of sound effect

There are four commonly used terms to describe sound fx. These are:

- *Spot fx/hard fx.* This term covers individual effects that are related to a single source (e.g. a plane, a door, a dishwasher, etc.). They are most commonly laid in sync with a specific action onscreen, but spot fx can also be laid to suggest out-of-shot action. Spot fx can be used in a literal way, the usual description of this being 'see a dog, hear a dog'. However, they can also be used in a non-literal way, when an effect is used for a purpose that is not related to its original source. For example, explosions are often used to add weight to falling buildings and breaking tidal waves, as well as to cover actual explosions. Spot fx can be used singly or multilayered, and may be cut or processed to change pitch, speed and eq.
- *Ambience/atmosphere fx.* This term covers non-synchronous fx, which are laid to create presence or ambience within a scene to place it in a specific location (e.g. traffic, birds, city skyline). Atmospheric tracks, or atmos, are usually laid in stereo, and when sent to the surround channels in the mix, go some way to creating a spatial effect which mirrors how sound 'surrounds' us in reality. Atmospheres are significant in setting the mood of a scene, and can be used to signify a change in location or a return to a location.
- *Foley fx.* These are sounds recorded in the studio directly to picture. Foley adds richness and realism to a soundtrack. There are three main types of foley fx – *moves*, which reproduce the body and clothing movements of the actor; *footsteps*, which recreate each footstep in the film, on the appropriate surface, in sync with the actor; and *specifics* – sounds that work alone as spot effects, or which sweeten spot effects laid by the fx editor, such as a cup being put down or a fight sequence where spot fx and foley combine to give a better sound. Foley is crucial in scenes that have been ADR'd, as the small movements that one would *expect* to hear need to be recreated to play over the clean dialogue. Foley is also an important element of the M&E mix, again replacing movement that is lost once the sync dialogues are subtracted from the mix.
- *Sound design.* This is an American feature-film term first used in *Apocalypse Now* and best illustrated in the rotor sounds made by the choppers. It is now used to describe the overall design of a soundtrack. In Britain, the term is used to describe the origination of specially designed sounds, and a sound editor will be more likely to incorporate *elements* of sound design into the tracklay. Audio can be manipulated through the use of samplers and plug-ins to create sounds that convey a degree of other-worldliness, or stylized mood. This term also covers the creation of 'new' sounds (e.g. futuristic sounds for a sci-fi movie). Sound design can be composed of layered fx which are manipulated to create sounds that are far removed from the original elements.

Planning the tracklay

Production requirements

The way in which a project is tracklaid depends very much on what type of production it is. Many types of TV production, such as serial dramas and 'lifestyle' programmes, will be simply but effectively tracklaid with the addition of a stereo atmos and a few spot fx, which are likely to be literal and picture led, such as doors, cars, etc. These fx will be tracklaid in the same session as the mix, often to a very tight schedule. TV programmes with a high degree of audience participation, such as sitcoms and quiz shows, will also require additional audience reactions to be tracklaid in the mix. In documentary/factual programming, the focus is likely to be on the use of actuality sound, to which spot fx and atmos may be added. Natural history documentaries, which are often composed largely of mute footage, are almost entirely created using atmospheres and foley fx, with relatively few spot fx. Commercials, which are usually dialogue and music driven, are usually not long enough to establish an atmosphere, but may contain a few spot fx to highlight certain key moments.

Tracklaying for feature films and high-end TV drama (usually shot on film) is a more complex process, where the requirement is to create a detailed soundtrack over a relatively long schedule. On location, the priority will always be to record clear and usable dialogue, which means that every effort will be made to minimize any other type of noise on the recording. However strong the narrative, the 'dialogue-only' track will lack impact and realism prior to the fx tracklay. (To counter this, some picture editors lay 'temp fx' in the cutting room for viewing purposes.) Atmos tracks and spot fx from a number of sources will be cut and laid over a large number of tracks, covering every action onscreen. In feature work, fx recording sessions often take place to create new sounds especially for the production, and there is often more time available to experiment with sound design.

It is important for the fx editor to know where the director plans to use music, and tracklay accordingly. Sound design in particular can often work in a very similar way to music in terms of mood and frequency, and any potential clashes should be sorted out before the mix.

Delivery requirements

Most high-end TV dramas and features require a full Music and Effects mix (M&E) to be made as part of the delivery requirements. This means that the effects editor should tracklay *all* action and supervise a foley recording session in which *all* moves are covered. The end result should be that, once the dialogue track has been subtracted from the full mix, there is a complete effects track that will be used to make foreign language versions later on. Other types of programme will merely require that the dialogue is subtracted from the full mix.

Playback medium

TV mixes have a narrow dynamic range and spatial spread, which means that the fx and dialogue are 'close together' and therefore more likely to conflict – where conflict does occur, the dialogue will always be prioritized over the effects, which may be thinned out or dropped in level. It is therefore a good idea to tracklay with the guide dialogue accessible, so that effects are placed within the scene to their best possible advantage. Projects that are being mixed for theatrical release can accommodate a very full and detailed tracklay without making the dialogue unintelligible. The high dynamic range and wide stereo/surround sound field means that sound fx may be placed away from the dialogue, which generally occupies the centre channel. When planning a track in 5.1 surround, some thought should be given to laying specific fx that will be sent to the surrounds channels and the subwoofer.

Sourcing sound effects

Finding the right sound effect for the job is a crucial part of the tracklay, and the editor can acquire sound effects from a number of sources.

Sync fx

These are effects which were recorded on the camera takes. Some may be used in the mix, others may be replaced in post production. Ideally, the fx should be split onto a separate track by the dialogue editor, so that they can be used in the mix as spot fx and used for the M&E, or dropped entirely. Sometimes it is important to make a sync effect work because the sound cannot easily be sourced in any other way (e.g. a period bell pull). It may be that the real acoustic on an effect is better than that which can be created in the studio, and so the sync effect is preferred to a library alternative.

Wildtrack fx

This term covers any atmos or spot effect that has been recorded on location without picture (or 'wild'). Wildtracks are especially useful when they cover effects that are difficult to source in any other way (e.g. an unusual car or a difficult location, such as the Antarctic). Crowd wildtracks are especially valuable for scenes where the presence of specific groups of people has to be created. If there is an opportunity, it is a good idea to talk to the sound recordist/production mixer about wildtracks in advance of the shoot. It's a lot harder to recreate the effect of 300 women in a Bingo hall in Liverpool after the event than it is to record it at the time!

Effects libraries

These are generally the first (and often only) port of call in any tracklay. General libraries from companies such as Sound Ideas and Hollywood Edge are organized into subject areas (e.g. sport, human, domestic, etc.) to make it easier to find a specific effect.

The best libraries are continually adding to their collections, and updating them to include new sounds. There are also a number of smaller libraries that specialize in a particular type of sound, such as 'Cartoon Trax' from Hollywood Edge, or 'Open and Close' from Sound Ideas, which is a five-CD collection of various doors doing exactly that. Library fx are generally clean close recordings. Occasionally fx have been processed before mastering, and some libraries offer variations of an effect in terms of reverb and perspective.

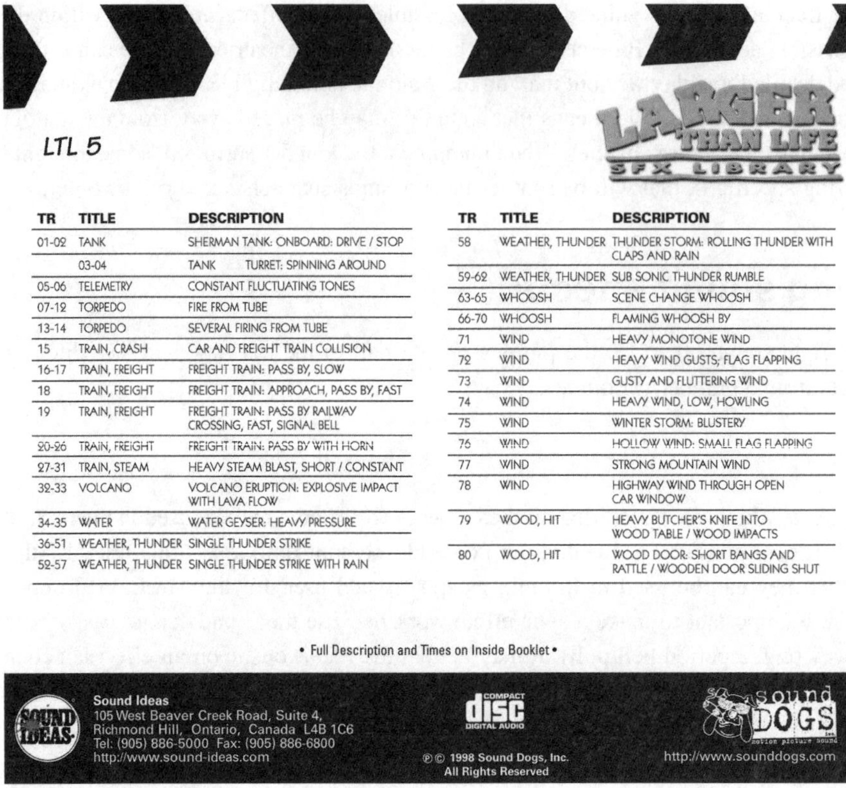

TR	TITLE	DESCRIPTION
01-02	TANK	SHERMAN TANK: ONBOARD: DRIVE / STOP
	03-04	TANK TURRET: SPINNING AROUND
05-06	TELEMETRY	CONSTANT FLUCTUATING TONES
07-12	TORPEDO	FIRE FROM TUBE
13-14	TORPEDO	SEVERAL FIRING FROM TUBE
15	TRAIN, CRASH	CAR AND FREIGHT TRAIN COLLISION
16-17	TRAIN, FREIGHT	FREIGHT TRAIN: PASS BY, SLOW
18	TRAIN, FREIGHT	FREIGHT TRAIN: APPROACH, PASS BY, FAST
19	TRAIN, FREIGHT	FREIGHT TRAIN: PASS BY RAILWAY CROSSING, FAST, SIGNAL BELL
20-26	TRAIN, FREIGHT	FREIGHT TRAIN: PASS BY WITH HORN
27-31	TRAIN, STEAM	HEAVY STEAM BLAST, SHORT / CONSTANT
32-33	VOLCANO	VOLCANO ERUPTION: EXPLOSIVE IMPACT WITH LAVA FLOW
34-35	WATER	WATER GEYSER: HEAVY PRESSURE
36-51	WEATHER, THUNDER	SINGLE THUNDER STRIKE
52-57	WEATHER, THUNDER	SINGLE THUNDER STRIKE WITH RAIN
58	WEATHER, THUNDER	THUNDER STORM: ROLLING THUNDER WITH CLAPS AND RAIN
59-62	WEATHER, THUNDER	SUB SONIC THUNDER RUMBLE
63-65	WHOOSH	SCENE CHANGE WHOOSH
66-70	WHOOSH	FLAMING WHOOSH BY
71	WIND	HEAVY MONOTONE WIND
72	WIND	HEAVY WIND GUSTS; FLAG FLAPPING
73	WIND	GUSTY AND FLUTTERING WIND
74	WIND	HEAVY WIND, LOW, HOWLING
75	WIND	WINTER STORM: BLUSTERY
76	WIND	HOLLOW WIND: SMALL FLAG FLAPPING
77	WIND	STRONG MOUNTAIN WIND
78	WIND	HIGHWAY WIND THROUGH OPEN CAR WINDOW
79	WOOD, HIT	HEAVY BUTCHER'S KNIFE INTO WOOD TABLE / WOOD IMPACTS
80	WOOD, HIT	WOOD DOOR: SHORT BANGS AND RATTLE / WOODEN DOOR SLIDING SHUT

• Full Description and Times on Inside Booklet •

Sound Ideas
105 West Beaver Creek Road, Suite 4,
Richmond Hill, Ontario, Canada L4B 1C6
Tel: (905) 886-5000 Fax: (905) 886-6800
http://www.sound-ideas.com

COMPACT disc DIGITAL AUDIO

℗ © 1998 Sound Dogs, Inc.
All Rights Reserved

sound DOGS motion picture sound
http://www.sounddogs.com

Figure 12.1 Effects disc from the 'Larger than Life' library (courtesy of Sound Ideas).

Libraries can be accessed in a number of ways:

- *CD*. Libraries are generally sold as a collection of CDs, which are then licensed to the purchaser for future use on any production. Searching through library catalogues and manually loading fx from a CD is now a relatively slow way to work. Unfortunately, CDs are also liable to handling damage and are expensive to replace. It is also true to say that the CD you want is often the one that's missing! A jukebox system goes some way to avoiding these problems, but a better solution is to hold the fx on a server.
- *Server systems*. The availability of increased hard disk capacity at lower prices has led to the development of systems which offer an intelligent search engine, coupled with instant access to sound effects from a central server. One such system is 'SoundBasket'. The system consists of three main elements connected by a standard Ethernet network – the *server*, which runs SoundBasket's database of effects; the *client terminals*, located in each studio or prep room, which run the search engine interface and replay the effects; and the centralized *library* of digitized sound files. Whenever a word or phrase is entered into the 'keywords' box, a complex algorithm extracts the root meaning of the entry, adding or subtracting standard or non-standard plurals, adverbs, synonyms, spelling alternatives (such as tire/tyre) and even common misspellings. Once the search results are displayed, the user can browse the list either one effect at a time, page by page, CD by

CD or library by library. Any sound can be replayed at full, uncompressed quality across the network through the inputs of the digital audio workstation, or drag and dropped as a WAV file. A system such as this can store and manage almost all the major commercial libraries currently available.

- *The Internet.* There are a number of sound libraries available on-line – one of the most comprehensive is Sounddogs.com, from which sound fx can be paid for individually on a buyout basis. Effects are arranged in categories and each effect comes with a brief description and its duration. Effects can be previewed instantly at a low bit rate and once an effect has been chosen, the buyer can select parameters such as sample rate and mono/stereo. File formats available are AIFF for Mac platforms, WAV for PC platforms, or MP3. After payment has been made, effects will be uploaded to an FTP site, from where they can be downloaded almost instantly, or can be delivered on CD. This is a useful resource for locating those difficult-to-find fx which may not be covered by the editor's own library.

- *CD-ROM/DVD-ROM.* These libraries are sold on the same basis as a CD library, but contain fx in WAV or AIFF format. Files are usually imported by 'dragging and dropping', which eliminates the need for real-time transfers. CD-ROM and DVD-ROM are ideal formats for libraries mastered in 5.1 and there is an increasing number of these available.

Figure 12.2 A sound effects database and server system: SoundBasket (courtesy of Raw Materials Software).

Recording custom fx

Some fx editors like to record their own atmospheres and spot fx, for a number of reasons. It may be that an effect is not available on any library, or not one which could be covered in the foley session (e.g. a rare vintage car). In this case, the editor might organize a quiet location and record the effects needed, most often to MiniDisc or DAT. For some editors, this is a way of creating a unique set of sounds for a production, rather than relying on library effects which, it may be felt, have been over-used on productions elsewhere. When recording atmospheres it is useful to record long tracks that can be laid over the full length of a scene without the need to loop.

Starting the edit

However detailed the tracklay is likely to be, it is a good idea to begin by searching for the majority of fx that will be needed during the edit, and recording these into the DAW. Grouping the fx by categorizing similar sounds will make it much easier to locate the desired effect later on. Once this stage is complete, the editor will know what additional fx will need to be covered by the foley session. At the very least, this might mean spending half an hour in a booth during the tracklay/mix of a documentary and creating some footsteps that are conspicuously absent on the sync track. On a more complex project, the fx editor will brief the foley editor in a detailed spotting session about the exact requirements for each scene (see 'Spotting the foley' below).

When thinking about the effects that could be needed for a scene, it is important to remember that sound effects can work on a number of levels. The most obvious use of effects is where they are required to match action on the screen in a literal way. However, sound effects can also be used in a *contextual* way. Here the effect chosen may comment or add to the scene, suggesting something that would not otherwise exist. Wind, for example, is often used to suggest bleakness and isolation in scenes where it is not literally needed. Sound effects can also be used in a *descriptive* way. Cicadas laid on a summer scene automatically suggest 'heat', even though they may not have been present at the original location.

The way in which a tracklay is approached depends on the style of the individual editor. Time permitting, it is always better to supply more detail than may actually be required (particularly on feature films), as the fx can be 'thinned out' at the mix. For a film drama or feature, a general approach may be to lay two or three atmos tracks on a first pass, then all picture- and plot-led spot fx on a second pass. Finally 'sweeteners' and extra touches can be added on the third pass. Figure 12.3 shows a contemporary battle scene followed by a domestic interior scene, which is tracklaid to be mixed in 5.1.

● *First pass.* Lay a skyline atmos with no distinguishing features, which may be sent to surrounds to create an illusion of spatial spread. Lay two atmospheres with detail that will be sent to the left, centre and right channels. In the interior scene, the sparrow and traffic atmospheres will be sent to the left/centre/right channels, and the large room atmos will be sent to the surrounds, creating a sense of the space in which the scene is taking place.

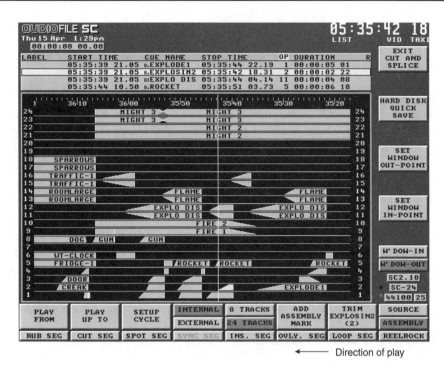

 — Direction of play

Figure 12.3 An effects tracklay (courtesy of AMS Neve/Jeremy Price).

● *Second pass.* Lay the plot/picture-specific spots required. There is a door in the interior scene. In the war scene, there are close-up explosions in mono and stereo, plus the fire tracks laid in mono on tracks 9 and 10. Although these two tracks look as though they're doing the same job, they are laid to be separately panned hard left and right in the mix.

● *Third pass.* Lay sweeteners that will enhance atmospheres and spot fx. Choose effects with appropriate natural perspective for distant action to create depth. Lay some spots that will be panned left and right. Distant machine-gun fire and explosions have been added to the war scene, which add realism and depth, and a flame sweetener added to one of the explosions. A distant dog has been added to the interior scene to add depth to the backgrounds, and a door squeak is laid to enhance the door open. All explosions have a stereo component added to the mono elements.

In this scene, body movements, footsteps and gun handling will be covered by the foley editor.

Tracklaying for the surrounds

Where a project is to be mixed in 5.1, sound fx can be placed accurately around the sound field by the sound mixer because the system has a good degree of separation between the speakers. Where a 5.1 project is converted to an LCRS mix, effects can still be positioned around the sound field, although the separation between the speakers isn't as distinct. In either case, it is more effective to lay fx tracks specifically for the surrounds.

The degree to which effects are sent to the surrounds is a stylistic device which some mixers make more obvious use of than others. Action films generally present more opportunities to spread fx into the surrounds than dialogue-led films. However, it may be that, in the battle scene described above, there is sufficient time to prelay a bomb whistle over the cut into the establishing shot, which, when mixed, will fly from the rear speakers to the front over the heads of the audience, and still remain clear of the incoming dialogue. Most effects spread into the surrounds will be atmospheres. When selecting an appropriate atmosphere, it helps to choose tracks that do not have prominent features within them, as this may result in the audience being suddenly distracted by a woodpecker, for example, that has just appeared at the back of the cinema! The atmospheres selected may be different to the atmospheres laid for the front, or the same atmospheres can be used for front and surrounds if appropriate. For example, in a dockside scene, the editor might tracklay stereo fx comprising seagulls, waterlap and a skyline atmos. In this case, the skyline would be a suitable effect to send to the surrounds, as well as centre, left and right channels. Using either of the other tracks for the surrounds would place the audience *in* the water or surround them with seagulls, which would distract from the onscreen action! In this case the skyline should be laid up twice, using a different start point within the cue, so that each track is out of time with the other, and no audible repeat occurs.

Tracklaying for the subs

The subwoofer channel of a 5.1 system carries only extremely low frequency notes, which are particularly effective in theatrical screenings. When a 5.1 mix is converted to an LCRS mix, the sound mixer may choose to include the low-frequency effects carried in the 5.1 channel, but will omit them when making the TV version of the same mix. So, for surround mixes, it is useful to lay effects specifically for the 'subs' to make the most of the dynamic range offered by surround systems. However, this is not

Figure 12.4 Waves' Maxxbass plug-in.

the case where a mix is intended only for analogue TV broadcast, as these low-frequency elements will fall outside its relatively narrow bandwidths.

Low rumbles, bass thuds and effects with a large bass component can be laid to create high impact effects or feelings of unease in the audience, who will not only be able to hear the effect but feel it bodily as well (e.g. the vibration of a train passing close by). There are a variety of subsonic rumbles and impacts, or 'sweeteners', available in fx libraries that can be laid underneath other fx to add weight and drama. Alternatively, the effect can be processed through a plug-in such as MaxxBass or an external effects unit to emphasize the bass elements, or sent through a harmonizer to lower the pitch between one and three octaves. The return will have sufficient low frequencies to register in the subwoofers.

Sound effects editing techniques

There are a number of techniques that can be used by the editor to improve a tracklay and enable it to be mixed more easily:

- Tracklay with level faders – regulating the relative levels between fx means that the mixer will have only small level adjustments to make, which will speed things up during the mix.
- Listen at a reasonable level – especially if tracklaying for theatrical release, as many effects sound very different at low levels.
- Use the establishing shot of a scene to establish interesting backgrounds, so that the viewer has registered the atmosphere before the (louder) dialogue begins.
- Build the atmos either side of a scene change, as this will signal the change in location to the audience. If the dialogue for a scene is prelayed, cut the atmospheres to match.
- If it is necessary to loop an atmosphere, make the loop as long as possible to avoid audible repeats.
- Atmospheres are laid in stereo unless related to a single source such as a clock tick, which can be laid in mono. Spot fx are mostly laid in mono. However, if the effect has a natural reverb, such as an explosion or a door slam, then it is better to tracklay it in stereo. If stereo spots are used, ensure that the stereo 'movement' travels in the same direction as the picture.
- It's not always necessary to be literal in choosing fx – the perfect track of seagulls flying overhead for a sunny beach scene may have been recorded at a landfill site!
- Sometimes it's better to not to replicate how things sound in reality – they may sound fairly flat and uninteresting! A crane will be much more interesting with the addition of some metallic creaks and servo motors.
- Spot fx should be laid frame accurately, using the production sound as a guide if the frame cannot be clearly seen visually (e.g. a door closing).
- In-shot effects can be very detailed and composed from a number of separate source effects to exactly match the onscreen action and timing. The effect of someone opening a door may be composed of the following elements: door handle turn, latch click, door creak and door handle return. Conversely, the same sound out of shot may only need to be a single effect (see Figure 12.5).

Figure 12.5 Edit detail: door opening created using several effects (courtesy of AMS Neve, adapted by Hilary Wyatt).

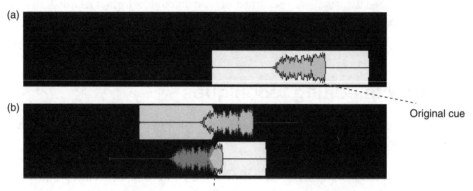

Figure 12.6 Edit detail: shortening an effect by cutting out the middle (a) Original sound effect. (b) Shortened edited version smoothed with crossfade (courtesy of AMS Neve, adapted by Hilary Wyatt).

● Where an effect tails off, don't cut it short – let it decay naturally. If an effect is too long (e.g. applause), it is sometimes better to shorten it by making a cut in the middle of the effect and letting it finish naturally (see Figure 12.6).
● The speed of an effect can be changed by flexing and also by cutting into it (see Figure 12.7). A fast car pass can be made even faster by cutting off the long approach and away, and putting steep ramps on the head and tail of the part of the effect where the car actually passes.
● The weight of an effect can be changed by adding or removing bass frequencies. Footsteps which sound inappropriately heavy can be made to suit a lighter person by rolling off some of the bass.
● Effects that do not work by themselves can be layered to achieve the right sound. A stab into flesh may consist of a body thud, a wet squelch, a knife movement and a bone crunch. Laying up a brake

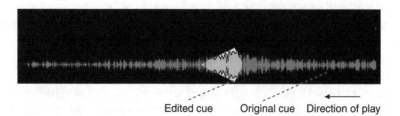

Figure 12.7 Edit detail: Changing the speed of a car pass effect by cutting and crossfading (courtesy of AMS Neve, adapted by Hilary Wyatt).

squeal, a tyre crunch on gravel and a handbrake will add definition and depth to the effect of a car arriving and stopping.

Sound effects plug-ins

Plug-ins are self-contained software modules that can be added to the audio chain and which will emulate a huge number of traditional outboard processors, both in function and appearance. The advantages of plug-ins are a saving in cost and space: software upgrades are available over the Internet, and multiple plug-ins can be chained together as required. The disadvantages are that plug-ins can vary hugely in quality and, once activated, use up processing power even if audio is not being processed. Most plug-ins are written to be used across Mac and PC platforms, and can offer functions such as reverb, eq, reversing, time stretching, phase reversing and time-based effects such as flanging, Doppler, phasing, chorus, etc.

Plug-ins also emulate a wide range of synthesizers, virtual instruments and samplers, which, whilst designed for music applications, can also be used for sound design purposes. Plug-ins are non-destructive unless a clip is rendered to create a new clip – in this case it is important to keep a copy of the original. Plug-ins can be routed in the same way as a hardware equivalent: the signal path is sent and returned in the usual manner via the host software's mixing desk, and plug-ins can be automated within the mixing environment.

Figure 12.8 Waves' Doppler plug-in.

Samplers and synthesizers

A sampler can play back a sound effect at varying pitches via a MIDI keyboard. An effect can also be assigned to each key so that the editor can 'play' effects into a scene using the keyboard. This is a quick method of tracklaying stock fx into multiple episodes of a game show, for example, where spot fx and audience reactions will be used repeatedly in each episode. Synthesizers (as a plug-in or actual hardware) can be used to create synth tones and effects that can be incorporated into the tracklay, and are particularly useful in manipulating sounds to create elements of sound design.

Presenting the tracks for the mix

The fx tracklay should be arranged across the tracks in the most manageable way for the sound mixer, who will be mixing the fx shortly after hearing them for the first time. On a project with a small number of tracks, the mix will take place in one pass rather than several premixes. Here the track allocation may be as follows:

Tracks 1–4   Dialogue
Tracks 5–8   Spots
Tracks 9–12   Atmos
Tracks 13–16   Music

On a larger project, where the effects will be separately premixed, tracks should be separated into atmospheres and spot fx, with tracks continuously assigned to stereo and mono respectively.

Figure 12.3 shows a 24-track edit that is organized in a way which makes it as easy as possible for the mixer to locate fx within the tracks and understand the editor's intentions.

- All spots which work together to form a composite effect have been placed on adjacent tracks where possible, apart from any stereo components, which have been placed on the higher tracks (e.g. the door is made of four elements on tracks 1–4).
- The tracklay is split, with mono spots/atmospheres occupying tracks 1–10 and the stereo spots/atmospheres occupying tracks 11–24.
- Any similar fx which will need the same eq setting are always placed on the same track (e.g. rocket fx on track 5).
- Distant spots are split off onto separate tracks, again to minimize changing fader settings in the mix (distant guns and dog on track 6).
- Stereo spots and atmospheres do not occupy the same tracks, as they will require very different fader settings in the mix.

Spotting the foley

The foley editor is briefed by the fx editor (and sometimes the dialogue editor) in a spotting session to establish areas where the foley will be particularly important and what foley is required by the director. It is important to specify any offscreen effects that may be needed, such as someone entering a room. Where it is not obvious, a decision will need to be made about the type of material or surface that will be used to create the foley effect. Where part of a scene is ADR going into sync sound, the foleys will need to match the sync fx exactly, so that the 'join' is not apparent. There may be a number of *specifics* that the fx editor cannot tracklay and these will have to be covered in foley.

The editor should then compile a list of cues, which may be just a written list of timecodes and a description of the effect needed; more formally, a cue sheet may be used, similar to that used for ADR. It is useful to let the foley studio know in advance whether any unusual props are going to be required, so that these can be sourced. The foley editor should supervise the session to ensure that all creative and technical requirements are met. (Recording foley is dealt with in Chapter 13.)

Editing the foley

However good a foley artist's timing, foley recordings will always need to be synced to the picture, and the best takes used selectively. If a foley effect is to work as part of an effect being laid by the fx editor, it is often easier for it to be cut and placed in the fx tracklay. Often in foley, less is more and lots of foley tracks running together can seem chaotic. The foley editor's skill is in achieving a high degree of naturalism whilst focusing attention on those sounds that are actually important. Tracks should be laid out in a way that is easiest for the mixer to manage. Maintain consistency in allocating specific tracks for feet, moves, etc. As with the dialogue and fx tracklay, all tracks need to be split for scene changes and perspective before being delivered to the mix.

13 | Post sync recording

Hilary Wyatt

As discussed in earlier chapters, most mixes will contain elements of studio-based recordings, such as ADR, foley and voice-over. With the exception of voice-over, such recordings should ideally be recorded in a similar acoustic environment to that of the original scene so as to be indistinguishable from the production sound. As this is generally not practicable, techniques need to be employed which ensure that the recording is of a quality that enables the mixer to recreate the appropriate acoustic in the mix using eq and artificial reverbs.

At the start of any session, the correct frame rate and sample rate for the job should be double checked, and a backup or 'snoop' DAT should be run throughout the session to eliminate the possibility of losing takes through system crashes, etc.

Recording foley

The studio

Facilities for recording foley can be, at their simplest, a variety of walking surfaces or 'pits' let into the floor panels of a voice-over booth. More complex foley should be recorded in an acoustically treated studio, where fx for both exterior and interior scenes can be recorded with equal success. Interior sounds are created by the sum of many different decay times reflecting from the various surfaces: exterior sounds tend to be characterized by a single reflection or 'bounce back'. The ideal space is therefore one which is neither 'live' nor anachoic, but which is treated to achieve smooth decay times across the frequency spectrum. The studio shown in Figure 13.1 is a large, near trapezoidal space which imparts little coloration to the foley, but within the studio area there are also areas that are dead and areas that are live should these be more appropriate to the recording.

Microphones

Generally, exterior scenes should be close-miked to eliminate roominess, with the mic approximately a metre away from the action. Interior scenes can be recorded with more 'air' around the foley, the mic

Figure 13.1 A foley studio with assorted props (courtesy of Peter Hoskins/Videosonics).

being positioned about 2–3 metres away from the action. Foleys are usually recorded in mono. The use of a cardioid mic such as a Neumann U89 tends to minimize studio 'hiss' and extraneous noise.

Organizing the session

A studio specializing in foley will have a huge range of props with which the artists can work. These might include doorframes, a water tank, stairs and car seats, as well as a collection of smaller items such as suitcases, crockery, shoes, various materials, etc. The foley team consists of one or two foley artists (two artists speed the process up considerably) and the foley recordist.

Using the cues provided by the foley editor, separate 'passes' of moves, footsteps and specifics are recorded for each scene. Such separation allows the recordist to mic each pass correctly and enables the foley editor to edit each track into sync. It also allows the mixer to adjust eq, reverb and level settings for each pass.

● Record the moves track first – this gives the foley artists a chance to familiarize themselves with the film before doing the most difficult pass, the footsteps.
● Listen back to the foley with the guide track – this will give an idea of how well the foley will 'bed in'.
● Try different mic positions or a different prop if it doesn't sound right – it's rare to capture a sound first take and be completely satisfied.

- There are many tricks of the trade used to create foley, such as placing a condom over a microphone and submerging it into a water tank to create underwater sounds. The artists themselves hold many secrets about how to create certain sounds, and will even conceal techniques from each other.
- To create the sound of a character walking from the back of a hall towards the camera, it's possible to use two mics (one tight and one held wide) and mix the ambience of the feet by fading the two mics in and out. However, most mixers prefer a balanced sound, leaving them the job of creating the ambience in the mix.
- The decision about mic placement can depend on the type of project being foleyed. 'Looser' recordings that impart a 'sync' feel can create a documentary/actuality style. Closer and richer recordings impart more weight and emphasis, giving the foley a 'Hollywood' feel.

Recording ADR

The studio

Facilities available for recording voices can be anything from a voice-over booth to a specially designed ADR studio. Recording ADR in a booth will cause a problem in the mix because the room reflections picked up in the recording will impart a 'boxy' feel that is impossible to remove. Using a microphone which is not reverberant field sensitive may counter this to some extent, but ideally ADR should be recorded in a large 'neutral' space that is neither acoustically dead nor reverberant. This is particularly important when recording shouted dialogue, or crowd groups, as high levels are more likely to trigger the natural acoustics of the room. In some instances, the recording room may have an acoustically dead 'tent' in which to perform exterior location ADR.

The actor should be able to sit or stand behind a lectern, which will help to avoid script rustle. He or she should have a clear view of the monitor on which loops are replayed and which also may show a visual cue. The monitor should not obscure the actor's line of sight towards the director and sound editor. If the studio has a separate control room, then a talkback system will be necessary to communicate with the actor between takes.

Microphones

The recordist should make an effort to match the mics used in an ADR session with those used on location, so that recordings are as consistent as possible with the sync track. Location sound is often recorded with a very directional boom mic (hypercardioid), which avoids picking up extraneous noise on set. Using this type of mic for ADR means that it does not need to be positioned very close to the actor, which is useful when recording indoor scenes. However, the mic should be moved in closer when recording outdoor scenes to avoid room coloration.

Personal mics (also known as radio or lavalier mics) are often used alongside the boom on location. If the ADR has to blend in with location sound recorded in this way, then the session should be recorded on both a boom and a personal mic. This type of mic is omnidirectional and prone to picking up noise such as clothing rustle, so takes recorded in this way need to be checked, particularly if the ADR performance requires physical movement on the part of the actor.

Where the location sound is quite roomy, it often pays to set up two booms, one close in and one further away. The mixer can then use some of the off-mic track to recreate the original acoustic of the scene and so obtain a better match.

If two mics are used, they can be recorded separately as a stereo-linked pair, which avoids editing the ADR twice and gives the mixer the option of using either side. Compression and eq should be used minimally as neither can be undone in the mix. For shouted performances, it is preferable to lower the pre-amp input level.

Cueing systems

To achieve the closest possible sync the actor needs to be given a frame-accurate cue for each loop. The ADR timecodes logged by the dialogue editor are fed into a cueing system that will trigger a cue for each loop. There are a number of systems available. The cue can consist of three beeps, a second apart, played into the actor's headphones. The fourth 'imaginary' beep is the cue point. A more accurate cueing system is the 'streamer', such as that made by CB Electronics shown in Figure 13.2. Here, a vertical white line travels across the screen from left to right at a predetermined speed. The start point is reached when the line reaches the right side of the screen.

Depending on the ability of the artist, it is occasionally preferable to use no cueing device whatsoever. This method involves the actor repeatedly hearing the line to be post synced and immediately

Figure 13.2 A streamer cueing system used for ADR (courtesy of CB Electronics).

afterwards performing the line without a picture reference. The disadvantage of this method is that each new take needs to be fitted manually into place to check sync.

Organizing the session

The attending sound editor should supply tapes of the locked picture cut and a set of prepared ADR cue sheets for each actor. During the session, it is the responsibility of the recordist to keep track of take numbers and note the selected takes. Sync and performance are the responsibility of the sound editor and director. It is best to try to load the cues for each actor prior to the session. Once the actors arrive, it is important to put them at their ease, especially if they are inexperienced at ADR. Ask them to read a few lines so that a good mic position can be established, and record levels set. The general method of working is to show the actor the line to be looped a couple of times, so that they can practice the rhythm, and then go for a take, continuing until one has been recorded that is satisfactory for both performance and sync. Actors can have a low feed of the sync track in their headphones as a guide during takes, although some actors don't like this.

- Equip the sound editor with a single headphone so that the new performance can be sync checked against the sync track during the take.
- Record one actor at a time, even if there are two actors in the scene – actors' ADR skills vary, and recording singly avoids awkward overlaps.
- If a *change* in performance is required, take the sync track feed out of the actor's headphone, as this will be confusing.
- Getting into a regular rhythm helps the actor concentrate and speeds up the session. An average session will complete about 10 loops per hour.
- Place selected takes on a specially assigned track as the session proceeds – this will speed up the fitting of the ADR at a later stage.
- On playback, dip out of the sync track for the looped dialogue and come back into the sync to check that the new line matches for performance and projection.
- Performances recorded whilst seated often sound too relaxed and lack energy. Whilst recording, ask the actor to stand – it will affect the timbre of the voice.
- Ask the actor to emulate physical movements that correspond to the scene to add energy to the performance (but ask them to keep their feet still). If an actor was lying down in the original scene, get them to do this in the session.
- At the end of the session, record a wildtrack of breaths and other non-verbal sounds from each actor – these will be useful to the editor later on.
- Where a whole scene has been ADR'd, play through the entire scene to check for continuity of performance, as well as checking individual lines as they are recorded.

Crowd recording

This is similar in technique to ADR recording, the requirement being to produce interior and exterior crowd fx that can be successfully blended into the sync track. Crowd is often recorded with quite large

Figure 13.3 A voice-over recording session, with ADR recording space to the left (courtesy of Peter Hoskins/Videosonics).

groups of actors, so mic techniques are somewhat different. When recording a group for an exterior scene, it is important to keep the group tight on the mic. Voices around the edge of the group may be a little off mic, and sound 'interior' as a result. When recording interior scenes it can be useful to use two mics, one close in front of the group and one positioned high and wide. The mixer will then be able to use some of the off-mic leg to add roominess, bedding the crowd more easily into the sync track. Crowd recorded in this way should be recorded twin track (but linked as a stereo pair for ease of editing). Where large crowds need to be created, record several takes, which can then be doubled up, but ensure that the group has changed position in between takes so that different voices predominate.

Voice-over recording

Voice-over has to sit on top of all other elements in the mix, and so the main requirement of a recording is tonal separation. Voice-over booths can be quite small, with a relatively dead acoustic. It is usual for the performer to sit at a table whilst recording. Figure 13.3 shows a voice-over booth, with a separate (and larger) ADR recording area to the left. The cueing system for voice-over is much simpler than for ADR: often a red cue light positioned in front of the artist indicates the start and end points of each cue. Voice-overs are generally recorded to picture, which results in a better performance and tighter timing.

A cardioid mic such as a Neumann U87 is ideal for voice-over work. This type of mic has a rounder and more intimate sound compared to boom or personal mics. It is not often used for ADR because its frequency response is such that it would be a challenge to match the ADR to the sync track. The artist should be close-miked: this brings the proximity effect into play (where the bass elements of the voice are emphasized), resulting in a warmer recording. Voice-over is usually heavily compressed in the final mix, so it is best to record the session with little eq and compression. As well as performance, each recording should be monitored for mic 'pops' or excessive sibilance, which can be unpleasantly emphasized by compression. The use of a pop shield reduces the chance of recording pops on plosive sounds, and angling the mic downwards towards the top of the palate can reduce sibilance.

Voice tracks for animation

Voices for dialogue-led animation are almost always recorded in a studio prior to any animation being shot at all. (A well-known exception to this is Nick Park's *Creature Comforts*, in which location interviews were used instead.) The selected takes from the session are edited together to produce an assembly that is close to the final running time. Once completed, the dialogue assembly is phonetically frame counted by a sound editor, who transfers the dialogue onto 'dope sheets'. These frame-accurate sheets are then used by the animators to create the 'lipsync'. Once the animation is shot, the picture and soundtrack should match up frame for frame.

It is important that dialogue is recorded 'flat' without coloration, so that artificial reverbs and eq can be added in the mix to create the appropriate interior/exterior acoustic for the eventual scene. This requires a recording space and mic set-up similar to that required for ADR. However, unlike ADR, it is often a good idea to record more than one actor at a time. This will help performance, as the actors will play off each other for delivery and timing. Due to the lack of pictures at this stage, it is a good idea to ensure that, whilst each actor is individually miked, they can all see each other and therefore pick up on any visual cues. Any accidental overlaps between actors should be re-recorded, as these could cause problems in editing later on.

ISDN (Integrated Switched Digital Network)

There are many times, particularly in film production, where it is not possible for the actor and director to be physically present in the same studio for an ADR/voice-over session. The solution is to link the two studios by a pair of standard ISDN telephone lines, down which can be sent any type of audio, albeit at a high cost per minute. Basic systems are based on two channels that carry 20 kHz bandwidth audio in real time, although Dolby Fax has an extended version that uses four channels operating at a much higher 256 kbits per second. Sony nine-pin machine control, timecode and audio can also be transmitted in either direction. This means that, assuming both studios are working to the same video standard, and both are working to the same picture reference, perfect sync can be obtained between the two.

Either studio can 'control' the session, but in practice it is more satisfactory for the controlling studio to be the one in which the artist is present. The receiving studio merely has to drop into record at the appropriate moment.

The disadvantage of ISDN sessions, apart from the cost, is that there is no face-to-face contact between the actor and director, which means that it is not really the best way to record a long session. Verbal contact, however, is maintained throughout down an open phone line.

When arranging an ISDN session in another country, it is important to specify:

● The video standard you are working to – this could mean the studio may have to hire in PAL or NTSC equipment, for example.
● Which studio is going to control the session.
● Check that both studios have received ADR cue sheets and guide pictures.
● How the ADR should be delivered and on what format. (The session should be recorded in both studios for safety.)

14 Preparing for the mix: music

Hilary Wyatt

Aims

Music is one area of audio post production that many directors and producers deliberate over most, and it can contribute a large part of the emotional content of a piece. Music is a hugely powerful tool in the hands of the filmmaker, regardless of the type of production. It is also a very versatile tool and can work on a number of levels within the soundtrack. Music editing can be as simple as laying the series opening title and end credits music, and locating a few short pieces of library music from CD to cover certain sequences. At the other extreme, it may involve a music editor working with a composer who will compose and orchestrate original music. From these examples, it will be clear that the production budget will determine what musical options are open to the director. Low-budget film and television work, and fast turnaround jobs such as TV promos, sports features, etc., will often be tracklaid using pre-recorded or *library* music. Another option is to create 'original' music by using a software program such as Apple's 'Soundtrack'. This enables the user to compose tracks by combining pre-recorded, royalty-free loops that can be edited to picture. Larger budget productions may commission music from a composer, who may use synthesizers if the budget will not stretch to a live recording session with professional musicians. For most feature-film work, the music budget will be large enough to accommodate live recording and pay for any songs that may also be used as part of the soundtrack.

- Music can enhance the mood of a scene.
- Music can help to pace a scene, particularly scenes that are actually cut to the music.
- Music can work as an atmosphere or ambience – for example, indigenous music is often used in travel documentaries.
- Music can assist in establishing time, location and period.
- Music can add dynamics to a mix if used selectively – wall-to-wall music loses its impact.
- Music can be sympathetic to a character's feelings within a scene or it can work contrapuntally against a scene to create narrative distance.
- The music can act as the director's voice, and through it he or she may direct the audience towards particular aspects of the story.
- Music can punctuate and heighten comedy.

- Music can develop progressively throughout a piece to help generate a sense of moving through the story.

Types of music

Score

Good score is the result of close collaboration between the director and the composer, who will create musical themes and textures that are appropriate to the style of the piece. The main advantage of scored music is that it is written to picture and will *underscore* changes in scene, action and mood. Score should also be written to 'sit' well around the dialogue. Another advantage of scored music is that it can impart a distinctive and original feel to the soundtrack, simply because it is used uniquely on the commissioning production and no other.

Source music

Source music refers to music that is mixed to appear as though it is emanating from an onscreen source, such as a radio or a TV, or a band within a scene. Source often works in a similar way to the atmosphere tracks in setting the scene in terms of time and location (e.g. Christmas tunes playing in a supermarket scene). Again, source music should be well edited to fit around dialogue and other elements in the soundtrack.

Soundtrack

Many features include a number of songs or soundtracks which work as score and which are often compiled into a CD on the film's release. The production will enter into negotiation over the use of each song individually, and the number of songs included in any one film is price dependent – costs can vary wildly per song depending on how well known the artist is. The advantage of using songs is that it enables the producers to focus the appeal of a film towards a specific group of people. The downside is that certain songs can suddenly seem to be overused on a number of TV and film projects, and what seemed like a good idea at the time may begin to look like a lack of originality. Stylistically, soundtracks are usually mixed at full volume with other elements in the mix lowered, or completely taken out, so conflict with dialogue and effects is not normally an issue.

Playback

Scenes that are shot to playback might include musicians in shot miming the playing of their instruments or dancers using the playback to keep in time. In either case, the same recording is played back on each take, so that the action maintains the same speed from take to take and can be successfully cut together later on in the edit. The disadvantage of this is that the location sound can only ever be a guide track, as it will contain both the playback and any footsteps/movements recorded on the take. Therefore, the clean recording that was used on set needs to be relaid in post production, and the musicians' moves and the dancers' feet recreated in foley.

Music and copyright

Music is classed as intellectual property and as such is subject to copyright law, which varies from country to country. However, international agreements exist which enable permission or clearance for use to be granted on a domestic or international basis, and most countries have one or more organizations that collect fees and distribute royalties on behalf of their members. In the UK, the MCPS (Mechanical Copyright Protection Society) deals with the rights to make and issue all formats of recorded music, and collect and distribute royalties (see 'Library music'). The PRS (Performing Rights Society) deals with the rights to public performances and broadcast music. It can charge a negotiated fee and issue a *licence* for a single use. However, the PRS can also issue a *blanket licence* to a production company, which will cover all uses of its members' musical works in broadcasts, cable programmes, on-line and telephone services, within an agreed period (usually annually). Internationally, performing rights societies include ASCAP, BMI and SESAC in the USA, APRA in Australia, and SOCAN in Canada.

Where a piece of music is not covered by an organization, the production will have to negotiate directly with the composer. Obtaining the rights for a particular piece is sometimes quite a complex task, and many TV companies will have a department that specializes in this. Companies also exist which specialize in music clearance. It is important to understand that the composer's rights (usually represented by the publisher) are legally separate from the performance rights (i.e. the rights of the person(s) who recorded it). Rights are usually granted on a non-exclusive basis for use on a specified production, and do not confer complete ownership of the piece upon the production. In practice, for each song or piece, the clearance process is as follows:

- Contact the composer(s), usually through the publisher(s) who will represent their interests. Where a song has been written by more than one composer (perhaps where there is a composer and lyricist, for example), they may have separate publishers, in which case both publishers need to be approached.
- Negotiate a fee with each publisher based on their percentage of ownership of the song and the way the song is to be used in the production. The fee will be based on how much of the song is used, how prominent it will be in the mix, and whether or not its use will involve a number of different media. A song used in a film, for example, will be reproduced on the film print, on DVD and VHS versions, and may also be included in a soundtrack compilation on CD. If permission for use is granted, a performance licence and a synchronization licence should be issued. The former allows the song to be used and the latter allows the song to be used in synchronization with picture.
- In the case of pre-recorded music, it is now necessary to contact the owners of the recording (usually a record label) and negotiate a second fee with them. Once this has been agreed, a master use licence should be issued, which allows the production to use that specific recording.

It is important to know that in situations where the performance rights are granted but composers' rights are not, the piece cannot be used. However, where the composers' rights have been obtained, but the performance rights have not, there are two possible options:

1. Find an alternative version of the piece for which performing rights may be granted. This can be a particularly successful option in the case of standard classical recordings that are often covered by music libraries (see below), where permission will always be granted.

SAMPLE MUSIC CUE SHEET

Series/Film Name:	The Big Show			Series/Film AKA:	Big Show
Episode Name	Let's Go			Episode AKA:	Go
Prod. #:	TBS 15		Episode #: 15	Show Duration	15:00
Original Airdate:	March 15, 2001			Total Music Length	01:07
Production Co./Contact Name:	Making Programmes Inc. Ms. P. Roducer 1234, Music Square Capital City, CA 12345-1245				

BI: Background Instrumental	VI: Visual Instrumental	EE: Logo
BV: Background Vocal	VV: Visual Vocal	
TO: Theme Open	TC: Theme Close	

Cue #	Title		%	Society	Usage	Timing
	Composer					
	Publisher					
001	The Big Show Opening Theme				TO	:32
	W	Jo Music	50%	PRS(UK)		
	W	Bill Wordy	50%	ASCAP		
	P	My Publishing Co.	50%	PRS(UK)		
	P	Jo Music	50%	ASCAP		
002	Big Show Closing Theme				TC	:35
	W	Jo Music	50%	PRS(UK)		
	W	Bill Wordy	25%	ASCAP		
	W	Fred Writer	25%	SESAC		
	P	My Publishing Co.	50%	PRS(UK)		
	P	Jo Music	25%	ASCAP		
	P	Fred's Music	25%	SESAC		

Figure 14.1 Sample music cue sheet (courtesy of Royalty Free TV).

2. In cases where the performance rights fee is so high that the production can't afford to use the recording, it may be cheaper to record a new version using session musicians. The production will then hold the rights to the performance and permission is no longer an issue.

It is equally important to know that there are occasions where the use of music is less than obvious and a breach of copyright can easily occur. When this happens, clearance will have to be obtained, which will incur extra cost. At worst, the music may need to be removed from the mix and a substitute found. Examples of this might include:

- Somebody whistling or singing a recognizable tune within a scene – if needed, it may be possible to get the actor to ADR something non-recognizable, and therefore copyright free, instead.
- An ice-cream van playing an audible tune within a scene – remove the tune from the tracks or pay for use.
- A mobile phone where the ringtone is a recognizable tune – replace with a ringtone from a copyright-free sound fx library or pay for use.
- Songs that are thought to be in common ownership but which are not, such as *Happy Birthday* – avoid including it in a scene or pay for use.

It is worth noting that projects (such as student films and shorts) which do not yet have a commercial distribution deal are usually offered very favourable rates for the use of music. In some instances, the production may be able to negotiate the use of a piece of music free of charge.

Music cue sheets

Once the project has been mixed, the production company must complete a *music cue sheet*. This document itemizes each cue, its duration within the mix (in minutes and seconds), the music title, composer and publisher, as well as the appropriate performing rights society. The cue sheet should also specify how the music was used in the final mix – for example, as background or featured music. The sheet is then used to report *actual* use of each music cue to the appropriate body and ensure the appropriate payment for that use. An example of a cue sheet is shown in Figure 14.1.

Planning the music

Music and the picture edit

On most productions, the editor and director will be thinking about the placing and style of music during picture editing. In some cases they may pre-select cleared tracks and actually cut the pictures using the music tempo as a guide. For fast turnaround projects, shows that may be tracklaid and mixed within an on-line NLE (such as sport inserts, news features, promos, etc.), the picture editor will select tracks with the director, edit them to the required length and incorporate them into the mix. Here, the music is often a combination of library music, for which permission does not have to be sought, and specially pre-written series theme music, which can be spotted throughout the programme and laid over the opening/closing credits. Where a project involves a complex music tracklay, it is more likely to be handed over to the sound department for a separate mix. Here, the editor may lay some music during the picture edit, which will be transferred to the dubbing mixer within the conform or OMF. Any music additions or changes can then be made by the mixer in the dub – most studios will have the main music libraries close at hand. A production that has an extended editing schedule is likely to have the option of commissioning a specially composed score. This is partly because there will be enough time within the schedule for the composer to write and orchestrate the music before the dub, and partly because the budget is likely to be large enough to cover the costs involved. As it is usual for the composer to work to the *locked* edit, most of their work takes place at the same time as the tracklaying, with delivery of the finished tracks often taking place just in time for the final mix. This means that the music is not available to the picture editor during the edit stage, so to get a feel of how finished sequences will look with sound, the editor and director may select music from a variety of sources and lay what is known as a *temp track*.

Temp tracks

Putting music to picture can considerably affect the emotional feel of a scene, and viewing a film with music is a very different experience to viewing the film with no music at all. A temp track enables the director to get an idea of the emotional impact the production will have once dubbed. A music track will be tested out against picture, discarded if not suitable and another one selected. This process will go on throughout the editing process using tracks from commercial CDs, film soundtracks and perhaps tracks from the selected composer, who might give the production music from other productions for which they have composed. Often this is music which will not, and cannot (because of budget limitations), be used in the final mix, but it will often set a style and tone which the composer will be required to emulate to some extent when writing the score. The drawback of doing this is that it is possible to get accustomed to hearing the temp tracks over a number of weeks or even months, so that when the final score does arrive, it appears to be a bit of a disappointment! On the other hand, temp tracks help the production formulate an idea of how much music is required and where it is likely to be used. Where feature films are tested during production to gauge audience response, a *temp mix* is made. Here, the temp music is crucial in helping the audience envisage how the finished film will sound, although the use of uncleared tracks for this purpose is legally a bit of a grey area.

Spotting the track

Once the picture edit is complete, a *spotting session* will be arranged to discuss music requirements for the production. This could be as simple as a quick briefing with the person responsible for finding tracks from the music library. A more detailed spotting session will involve discussions between the director, editor and composer about musical style and the type of instrumentation that would be suitable. Each piece of music, known as a *cue*, will be discussed at length in terms of its in and out points, duration and any specific *hit points* – moments in the music where a musical feature corresponds to an action onscreen. The director will need to convey to the composer what each cue needs to achieve within a scene (in terms of emotion, comedy, suspense, etc.) and indicate where any changes in mood or location should take place. Each cue should be given a reference number, which can then be used to distinguish each cue in subsequent discussions, and later on when the music is tracklaid or *laid up*. For example, cue 1M3 may refer to the third cue in Episode 1.

Delivery requirements

Many mixes for TV are required to be made in stereo. Indeed, for some factual formats, the music may well be the only stereo component in the mix. This means that all music sources should be stereo and tracklaid in stereo whether specially composed or from CD. Where a production is to be dubbed in 5.1, it is usual for the composed music to be delivered to the mix in 5.1. Once the final mix has taken place, it is usual for the delivery requirements to stipulate that:

- A copy of all score cues (in 5.1) is supplied on a format such as DA88.
- A stereo mixdown DAT be made of all original cues and source music cues as they appear in the mix, but without the accompanying fader moves.
- A stereo DAT of the complete versions of all original source music cues used in the production is supplied, together with the details of each copyright owner.

Playback medium

TV mixes have a narrow dynamic range and spatial spread, which means that the music and other sound elements are close together and therefore more likely to conflict – where this does occur, the dialogue will always be prioritized over the music, which may be dropped in level or dropped altogether. Some composers will add a large bass component to their music, so that it sounds big and impressive when played back for approval in the composer's studio – this will need to be rolled off below 20 Hz in the final mix. Theatrical mixes benefit from a wide stereo/surround sound field, which means that music can be placed around the dialogue (that generally occupies the centre channel). When mixing in 5.1 surround, all source tracks (from CD, for example) will be in stereo. In order to get the most out of a stereo track in 5.1, it may be necessary to apply eq, filters and delays to effectively spread the music into the surrounds and the subs.

Sourcing music

Library music

This is also known as *production music* and is specifically written to be used in commercials, film and TV production, as well as other audio-visual media. Music libraries are generally available on CD and CD-ROM, as well as on-line. Libraries do vary in the quality and range of music offered, but tracks are instantly available for use at a comparatively reasonable cost.

Paying for use

In the UK, once a *mechanical licence* has been obtained, the MCPS charges a fixed rate per 30 seconds of library music based on actual use. So, where the music cue sheet states that a track has been used for 32 seconds, the production must pay twice the appropriate rate. Rates depend on the media (feature films and commercials are charged the highest rate) and the intended distribution. TV rates, for example, increase depending on whether the production will be shown in the UK and Ireland, Europe or worldwide.

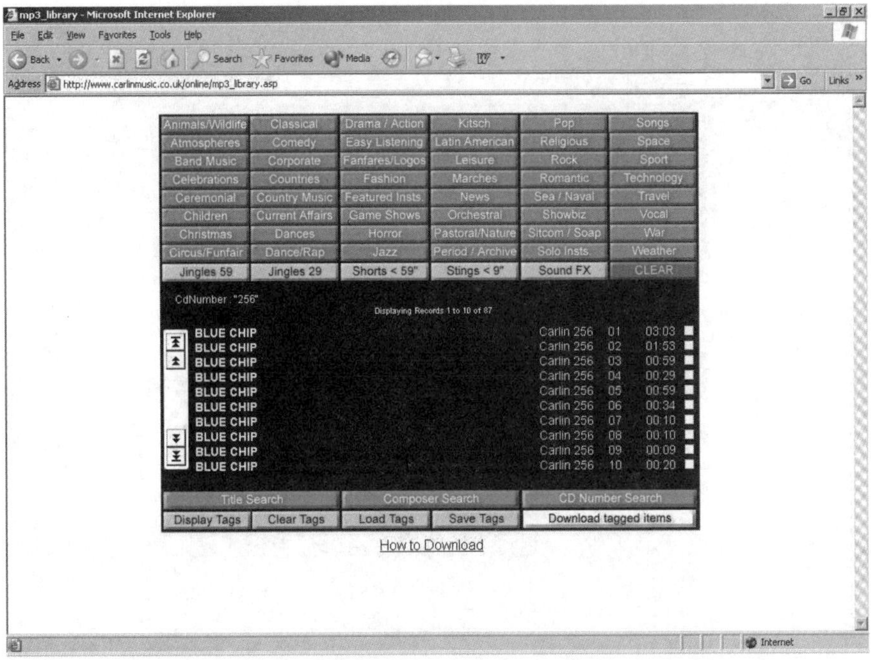

Figure 14.2 Library music search engine (courtesy of Carlin Production Music).

In the USA, mechanical licensing for production music is handled primarily by the Harry Fox Agency, as well as SESAC and AMRA. Once a licence has been issued, HFA charge a standard rate for songs less than 5 minutes in duration. For songs over 5 minutes in duration, a rate is charged per minute or fraction of a minute. These rates only cover productions intended for distribution inside the USA.

Some music libraries are available on a complete *buyout* basis, rather like fx libraries. This means that, once purchased, a track can be used repeatedly without additional payment.

● Each track is catalogued and described in terms of its tempo, musical style and whether or not it builds, as well as its instrumentation. This makes searching fairly straightforward and unsuitable tracks can be quickly discarded.
● Most libraries are equipped with a search engine into which keywords can be typed – such as 'exciting' + 'chase' + 'builds to climax'.
● Libraries will often do a search for a client themselves, once they have received a brief description of the music required, and come up with a number of suggestions which can be e-mailed or sent out on CD.
● Some libraries offer historical and archive music that sounds like the real thing – because it is! The De Wolfe library, for example, has a number of adventure/action/suspense tracks from the 1930s and 1940s.
● Libraries are often good at musical 'pastiche', which can be useful for animation and comedy.
● Libraries offer sound-alikes, which can be used when an original track is too expensive.

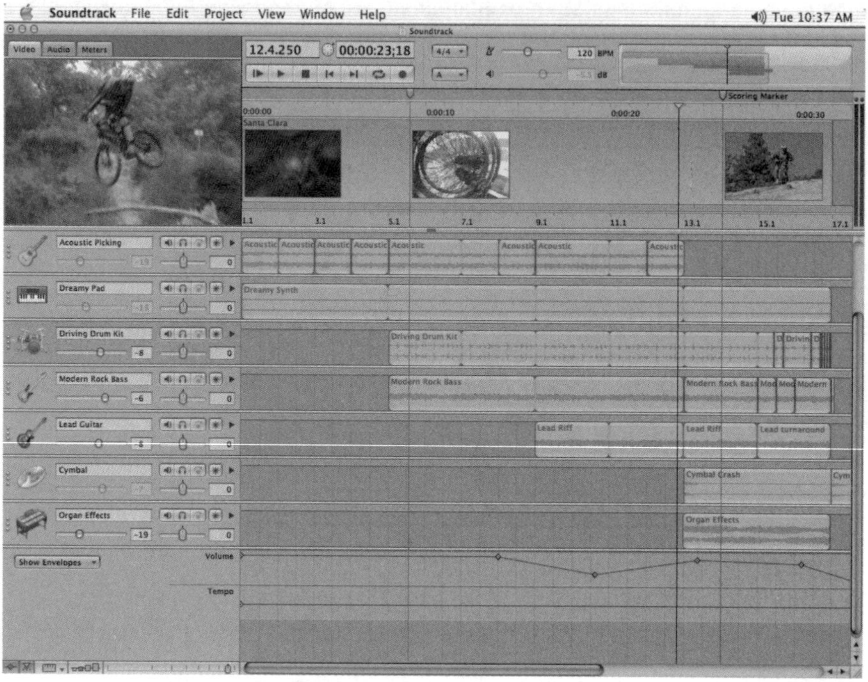

Figure 14.3 Apple Soundtrack playlist (courtesy of Apple).

- Libraries offer a production the possibility of using music recorded by live musicians – this is beyond the budget of many TV productions, even if they can afford to commission a score.
- Tracks are written in the knowledge that they will be edited to fit, and therefore will have a number of natural edit points.
- A track is often supplied in a variety of lengths, again to enable the track to be used a number of ways – for example, a 20-second commercial cut, a 40-second version and a full-length 2-minute track.
- Music does not need to be cleared prior to use, which is crucial for some productions.
- A single licence (from the MCPS in the UK) covers the composer's rights, recording rights, synchronization rights and the performer's rights – these would normally have to be obtained separately.

The disadvantages of library music are that it needs to be skilfully edited to fit both in terms of overall length and hit points. It is absolutely fine for many types of production, but will not be able to follow the emotional mood of a drama, for example, as well as scored music. Good library music is also bound to be used on other productions, which may detract from its appeal to a director and its effectiveness.

Music creation software

There a number of software programs available that allow the user to edit and combine a selection of pre-recorded copyright free musical loops. Apple 'Soundtrack', for example, is designed to be used by non-musicians and comes supplied with 4000 different loops, which can be edited to picture, and is imported as a QuickTime movie.

The software is also equipped with fully automated mixing and effects, such as eq, delays and reverb settings. Other sounds can be imported into the project in Acid, AIFF and WAV formats. Soundtrack will play back any number of instruments in sync with picture, and the tracks can be constantly modified until the final desired effect is achieved, at which point the mix can be exported into an external editing system using MIDI timecode to maintain sync. Another program, which does much the same thing, is Sonic Foundry's Acid. In either case it will take some time to produce music in this way, but it may be a good option for low-budget productions that wish to achieve a unique sounding track.

Composed music

Where a composer is hired to provide an original score for a production, the music can be composed and mixed using samplers and synthesizers, often in the composer's home studio. This will be the case for many TV productions. Alternatively, the score can be worked on in the composer's studio until a rough version of the composition has been completed using synthesizers. Once this version has received the director's approval, the music is re-recorded in a professional recording studio using live session musicians. For obvious reasons this option is only open to high-end productions that have a substantial music budget. The composer may be paid a fixed fee for the composition and orchestration of the music, or may agree to work on a *package deal* – here, the fixed fee must cover all expenses, including composition, musicians, studio time and stock. In either case, the production will generally hold complete ownership of the music and the composer will be paid a share of any future earnings in the form of a royalty payment. The music production process is as follows:

● The composer is chosen by the director and producer(s), usually on the basis of previous work done for other productions. Sometimes this may involve an 'audition' to see if their ideas for the music match the vision of the director.
● Once the picture is locked, the composer is given worktapes that must be made with the correct timecode start point and frame rate for the project. The worktapes should be made with burnt-in timecode and guide dialogue. The composer can start to work, matching the timing of scenes, specific hit points and the emotional movement of the action. A composer may take around 6 weeks to complete an average film score. As post production schedules are very often compressed, it is not at all unusual for a composer to start work before the picture edit is locked.
● The director and composer collaborate to create themes and textures with specific instrumentation using MIDI to maintain sync. These tracks may be mixed into a final score and delivered to the dub, or may be a rough visualization of how the music will look against picture after the live recording session. The composer may also collaborate with the sound effects editor to discuss areas where the music and effects might clash, or to discuss ways in which sound design and music may complement each other.
● The final stage is to create the mix master in a recording studio with a number of musicians and a conductor. This often takes place at the beginning of the main mix (in the dialogue premix week). The musicians will be given an audio click track in their headphones, which will indicate any

changes in tempo, and the conductor will conduct to picture. A visual streamer system may be used to indicate the start and end of each cue, or alert the conductor to an upcoming picture reference that may need to be musically scored. An average 3-hour session will yield between 8 and 10 minutes of finished music, maybe less.

● The best takes from the session will be assembled and mixed by a recording engineer to produce a mix master in the appropriate delivery format. This may take a similar amount of studio time to the record sessions themselves. Stereo mixes may simply be transferred to DAT, whereas multichannel mixes may be laid off to DA88 or exported as a file transfer. The music should be delivered to the final mix accompanied by a written list of cue names and their start times, in case any sync issues arise.

The role of the music editor

On larger productions with complex music requirements, a music editor is often employed specifically to assist the composer and liaise with production on music-related issues as the film is edited, track-laid and mixed.

● The editor may oversee playback music on set if needed, and organize playback tapes.
● The editor will assist in finding and cutting temp tracks if needed. The chosen tracks are cut and fitted to picture and may be used in test screenings during the editing process.

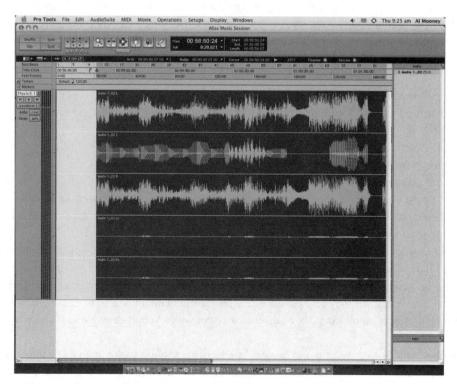

Figure 14.4 Importing a multichannel music file into Pro Tools (courtesy of Digidesign).

- Once the project is locked, the editor will attend the spotting session and make notes for each cue.
- He or she will organize and edit source music tracks not written by the composer, and assist in obtaining clearances.
- He or she may prepare click tracks and visual streamers using 'Auricle' and 'Cue' software.
- The editor will oversee the recording session, making notes on each take, noting each selected take and keeping timings for each cue, as well as a record of how much material has been covered in the session. He or she will also keep a record of anything that will need to be overdubbed – for example, a vocal track or a solo.
- He or she will attend the music premix and may ask for certain instruments to be mixed to a solo track. (For example, horns used in an important dialogue scene can be dropped down in the final mix if speech audibility becomes a problem.)
- The editor will lay up the music at the correct timecode positions and may recut the music to any subsequent picture changes. This must be done in a way that tries to preserve the composer's original intentions.
- He or she will attend the film mix, carry out any fixes that are required on the dub stage and generally look after the interests of the music.
- Finally, the editor will prepare the music cue sheets (see above), which are then given to the production company to pass on to the appropriate performing rights body.

Starting the edit

Loading the music

Once the music tracks have been found, or the score delivered, it can be imported into the DAW. Importing from a non-timecode source such as CD or CD-ROM is simple: the track is recorded in and placed on the timeline, where further editing can take place if necessary. Where the music has been written or edited to picture, the track must be imported with a sync plop or with SMPTE timecode in order to maintain sync reference. Multichannel mixes (where the music has been premixed in 5.1,

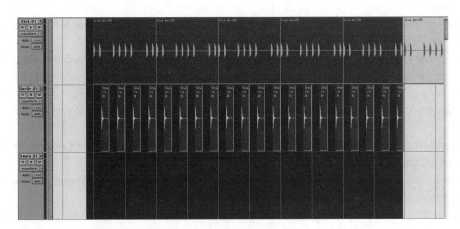

Figure 14.5 Editing music with markers (courtesy of Digidesign).

for example) may be played into the mix directly from a multitrack format such as a DA88 machine slaved to the studio controller. Alternatively, the music can be recorded into the DAW preferably via the digital inputs or imported as a file transfer. In either case the audio will appear in the DAW as six mono tracks, and these should be grouped together so that the channels remain in phase with each other during editing and resyncing.

Editing the music

In editing music to picture, it is important to be aware of the structure of the music and try to make edits that make musical sense. A starting point is to look at music in terms of beats, bars and phrases, which may make the natural edit points in a piece more apparent. It may be possible to set the time-line to count in beats and bars as well as timecode. Some editing systems offer software that can identify the beats within a cue, and place markers or even cuts on beats and phrases.

In using these to shorten a cue, for example, a cut can be made in a track that has been positioned at the desired start point. The end section should be moved onto another track and pulled up so that it finishes at the desired point. The resulting overlap can be scrubbed for a suitable crossover point, and a cut made on both tracks at the nearest marker. The two tracks can then be butted up against each other, and the join checked to ensure a good edit. Generally, however, more complex editing is intuitive rather than automatic and the track will need to be listened to many times and a number of edits tried out before a satisfactory cut is achieved.

Music editing techniques

There are a number of ways in which music edits can be improved:

- Think like a composer, not like a sound editor.
- Always cut on the beat! This will help to retain musical timing whilst editing, and the edit itself will be masked by the (louder) beat immediately after it (another example of temporal masking used to good effect).
- Make sure that edits and crossfades on the beat do not flam (i.e. a double beat is heard because the beats are out of sync).
- When a cut is made on the beat, double check that the composer hasn't preceded the beat slightly with another note. If this is the case, roll back the edit point to the head of the note – this will prevent the edit from sounding clipped.
- Always try to make musical sense – even at the expense of where musical features fall against picture.
- If this is absolutely impossible, a pragmatic approach is to hide the offending edit under a louder element in the mix, such as a sound effect.
- When shortening a cue, it is important to retain any transitional phrases between key changes. If these are removed, the key change will sound musically wrong.
- When laying source music as though coming from a radio/TV/hi-fi within a scene, it is conventional to omit the intro and choose a start point some way into the track – often the start of a verse or

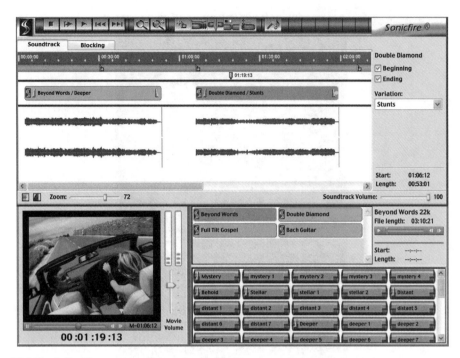

Figure 14.6 Sonicfire Pro.

phrase and preferably with no immediate key changes. This will make the cue seem naturally part of the scene and less premeditated. It is more important to choose a 'comfortable' in point for the music at the start of the scene than the end. The audience is much more likely to notice a jumpy incoming edit, whereas the out point will be perceived as part of the scene change, even if it occurs mid phrase.

● Where source music containing lyrics is to be placed under dialogue or voice-over, it is important to ensure that the selected track is edited and mixed in a way that does not compromise intelligibility. It is preferable to establish the vocals at full level long enough for the audience to register them before dipping the music to accommodate the dialogues. This ensures that the music has maximum impact, without being distracting.

● Where music is being edited for a film production, it is crucial to bear in mind that the sound on the final digital print will be 26 frames ahead of the picture. When laying a music cue near the head of a reel, the first 26 frames of picture at the head of each reel must be left completely free of music. Likewise, at the end of the reel, the last music cue must not be longer than the picture, but must end on or before the last frame of picture. This ensures that the music does not get clipped on the reel changeover when the film is projected in its entirety.

Plug-ins

There are software programs that attempt to automate music editing completely – one such program is 'Sonicfire Pro'. Pre-recorded production music can be auditioned from the search engine against

Figure 14.7 Pitch 'n Time plug-in (courtesy of Serato Audio Research).

picture that is imported as a QuickTime movie. The music itself has been pre-loaded in *blocks* that the program strings together to match the cue length determined by the user. The program simply ensures that a cue will start and end naturally: any intervening hit points cannot be accommodated. It is, however, a very quick way of fitting production music where general themes are required.

Organizational software used by music editors may include 'Cue', a program which is used for generating timing notes, and 'Auricle', a program which can control all timing parameters for a live recording session on a beat-by-beat basis, such as hit points, tempi, timings and streamers.

There are a number of plug-ins that enable the editor to adjust the pitch and tempo of a piece of music. 'Beat Detective' allows the user to adjust or clean up the tempo of a piece by *quantizing* the beats. 'Pitch 'n Time' allows the user to modify the tempo of a piece between 12.5 and 800 per cent of the original, and pitch shift by ±36 semitones. This software also has a variable mapping function, which means that both pitch and tempo can be adjusted within a piece to hit specific picture references. This may be particularly useful where a scene is recut and the score needs to be edited to fit. This type of software could equally be used by a sound designer to modify the pitch and tempo of sound fx within a scene.

Other music-orientated plug-ins perform noise reduction tasks. Recordings from old analogue formats such as 78 rpm records may be improved by using a number of noise reduction processors, such as the DeScratch, Dehiss and Eq combination manufactured by Cedar. Poorly recorded source tracks may benefit from being processed through a plug-in such as the BBE Sonic Maximizer, which can add crispness and clarity to an otherwise 'muddy' track. In addition to these specialist programs, there are a large number of plug-ins that offer eq, reverbs, delays, compressor and gating functions designed specifically for music applications. However, if the music is to be successfully balanced together with dialogue and effects in the final mix, many of these processing decisions are best left to the sound mixer, who will have an overall view of how the music should be processed in the context of the sound-track as a whole.

MIDI

MIDI (Musical Instrument Digital Interface) is a way of interfacing a computer to various MIDI devices, such as samplers and synthesizers, which are controlled by a keyboard. These devices are synchronized to the computer via a MIDI interface that also supports timecode, so that a sequence can be played back in sync with picture. The computer runs sequencing software (such as Cakewalk or Logic Audio) that allows the user to create a MIDI composition file. Each keyboard action is memorized by the sequencing software, so that each note (and its length) is replicated with precise timing on play-back. Many instrument parts can be played out simultaneously from a number of tracks into a MIDI output channel. The tracks themselves may be edited to correct timing, the duration of individual notes changed and moved, and tempo and rhythm adjustments made to a sequence. Some sound effects editors use MIDI keyboards to play effects into a scene. In this case, each sound effect is assigned to a specific key and played out against picture. It should be emphasized that MIDI timecode (sometimes referred to as MTC) does not carry actual audio, it consists only of timing information.

Presenting tracks to the mix

Source music should occupy specific tracks in the tracklay, as this music will need to be equalized and processed in a very different way to the score. Source music should be cut across several tracks for perspective – in the same way as the dialogues and atmosphere tracks – so that the appropriate eq and reverb can be applied in the mix.

Where music is recorded as a multitrack cue, a good split for an orchestral score might occupy up to 16 tracks as follows:

Tracks 1–6   Orchestra (Left, Centre, Right, Left Surround, Right Surround and Subs)
Tracks 7–9   Vocals (Left, Centre, Right)
Tracks 10–12   Synth, Rhythm (Left, Centre, Right)
Tracks 13–14   Solos
Tracks 15–16   Spare tracks for extra surrounds, or centre tracks if needed.

15 | Monitoring and the environment

Tim Amyes

Soundtracks are reproduced in various different acoustic environments: the home, the cinema, the conference hall. In an ideal world, all control rooms and listening rooms would be standardized to allow perfect recordings to be made. The Dolby organization provides standards and licences for film re-recording theatres, whereas for small monitoring and mixing spaces the THX organizations analysed the needs and produced their PM3 specifications (Professional Multichannel Mixing and Monitoring).

However, most audio mixing areas are not built to these ideal and exacting specifications, and since sound is very subjective, and no two people are likely to interpret sound in the same way, it is surprising that there is any similarity between recorded sounds at all!

A mixing room for monitoring sound must possess certain characteristics:

- The sound control room should be the definitive listening area for a production and reach the highest possible quality standards.
- The control room needs to be of reasonable size for multichannel sound, otherwise only one person may be able to sit in the correct listening position – a slight movement of the head will lead to a significant change in the stereo image. Far-field monitors with a larger sound field will allow more people to hear the image successfully.
- When listening at normal listening levels, within a digital environment, it shouldn't be possible to hear the background noise of the recording medium, unless there is a problem. This enables a check to be made if there is a build-up of noise through the system.

In the audio post production control room, decisions are made during the sound mix on:

- Aesthetic judgements of levels, perspective of sounds, fades, dissolves and cuts, and special effects.
- Correct synchronization, making sure there is the correct time relationship between spot effects, dialogue, music, etc.
- Technical quality concerning the acceptable frequency response, phase errors, and noise and distortion of the system.

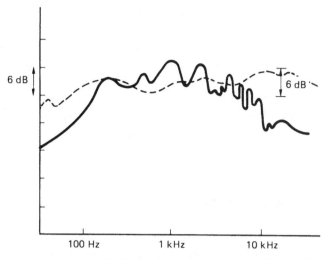

Figure 15.1 The audio responses of a studio monitoring loudspeaker (broken line) and a good-quality television receiver (solid line).

● Placement of the stereo image and surround sounds, making sure that the sound moves correctly in relationship to the pictures across the stereo field.
● If the production is to be transmitted on television, there needs to be stereo and surround sound compatibility with mono, making sure that the viewers in mono receive an intelligible signal.

Monitoring loudspeakers

The loudspeaker is the most important instrument in the audio post production studio; from it, everything is judged. It has been said that loudspeakers are windows through which we view sound images, and this is a fairly apt analogy. However, sounds are more difficult than pictures to interpret correctly; an experienced eye can easily check the colour balance of a colour photograph by comparing it with the original. Sound, however, is open to a more subjective interpretation, being influenced by acoustics, reverberation, loudspeaker distance and volume. All these affect the quality of sound we perceive through the window of our loudspeaker.

Loudspeakers are notoriously difficult to quantify. Terms such as harshness, definition, lightness, crispness and the more modern terms, translucence and sonic purity, are commonly used. But these different terms have different meanings for different people; multichannel sound adds yet one more dimension to the problem. There has been some attempt to provide a quality standard for loudspeaker design by the THX organization, who produce an approved list.

The highest quality speakers are designed for use in large studios. For editing and mixing in smaller environments, units are manufactured in smaller enclosures for 'near-field monitoring', but unfortunately they are unable to reach the standards of their larger brother. The overall frequency response of monitors is important, although many people think they have full range speakers; in fact, few monitors are capable of reaching the lower frequencies below 40 Hz. Surround sound demands lower frequency reproduction and separate subwoofers are often required. Interestingly, most home surround systems have low-frequency speakers that will pick this function up, but may not extend below 80 Hz in the main system. Surround systems are now defined by the number of discrete channels (between five and 10) and by the low-frequency channel. This gives a designation of 5.1, 7.1 or even 10.1. These digital systems have developed from the original analogue four-channel system devised by Dolby (SVA), having left, centre, right and surround signals. This is still offered in film release prints today.

Stereo and multichannel sound

Stereophonic sound conveys information about the location of sound sources, but this ability to provide directional information is only one of the advantages of stereophony. Stereo recordings provide an improvement in realism and clarity over mono recordings. They have a clearer distinction between direct and reflected sounds, which produces a spacious three-dimensional image with a better ambient field. Pictures can be powerfully reinforced by the use of stereo sound with better directional cues for off-stage action, and with sound effects that have increased depth and definition. The film industry was quick to see the advantages of 'stereo' in a large environment and has led the way in multichannel sound even before television became a mass medium.

Even television, with its small sound field, benefits from stereo sound. The positioning of sounds may not be practical, but the recreation of an acoustic can be used to advantage. In a multi-camera production it might be impractical to position the microphone correctly for each shot in a rapidly cutting sequence; however, a general stereo acoustic can give an impressive result in surround sound, particularly in sports and current affairs programmes.

Recording stereo sound and surround sound requires more equipment and more tracks. Post production time is increased and there are small increases in recording time on location or on the set. Thus, stereo productions are more expensive to produce both in terms of time and in equipment needed. It has been estimated that it takes 25 per cent more time to post produce a television programme in surround sound than mono. In the cinema, it has been the exhibitors who have resisted the change to multichannel sound, with its high costs of re-equipping their theatres.

The multichannel sound formats used in audio post production are:

● The simple left–right stereo format used in stereo television (two separate discrete channels).
● The cinema left, right and centre with surround format (LCRS), used in Dolby Stereo.
● The five-track surround formats used in film and the digital versatile disc, which have an additional low-frequency channel – the system is designated 5.1.
● The seven-track surround system of Sony Dynamic Digital Sound, which also includes a low-frequency channel, 7.1.

If stereo and surround recordings are to be satisfactorily monitored, the sound image reproduced needs to have a fixed relationship to the picture. It is important, therefore, that there are sufficient sources across the sound field. In television, where the picture image is small, a pair of speakers positioned either side of the screen produces a satisfactory image. In the cinema, where the width of the screen is perhaps 15 metres, at least three speakers are required to reproduce the sound satisfactorily, designated left, centre and right. In addition, there are surround channels within the theatre auditorium itself. The Dolby Stereo analogue format is of this type. However, although four channels are heard in the mix, only two separate channels are actually recorded on to the soundtrack.

The four channels are reduced to two by using a matrix. Matrixed systems are not entirely crosstalk free, so it is important that the sound mix is monitored through the matrix when it is being recorded. Through the matrix, the information on the left is fed directly to the left track (Lt). The information on the right is fed directly to the right track (Rt). The centre is fed at a reduced level to both, in phase. The surround is fed at a reduced level to both, out of phase. Sound from the screen speakers can be heard in the surround speakers. However, this effect can be all but eliminated by adding a delay line to the surround speakers, and restricting the frequency response of the system to 7 kHz. The brain then identifies the first place from which a sound is heard as the sound's origin, and mentally ignores other sources of the same sound arriving a fraction of a second later. The signal from the front speaker reaches the ear first, so the mind ignores the same signal from the surround (this is known as the Haas effect). We discuss the practical problems of mixing material to be encoded in Chapter 16.

With digital surround formats the channels are recorded separately, not matrixed. Five or more discrete channels are used, with one providing low-frequency enhancement. Multichannel discrete systems have none of the matrix and monitoring concerns of Dolby Stereo.

In domestic two-channel hi-fi stereo sound systems, it is generally recommended that the speakers are placed at angles of 60 degrees to the listener. Smaller angles make it difficult to judge the stability of the centre (or so-called phantom) image, and larger angles can remove the centre image almost completely (leaving 'a hole in the middle'). At angles of 60 degrees, the listener and speakers are equally and ideally spaced from each other, the listener sitting at the corners of an equilateral triangle. This system has been adapted for television, where there is the additional requirement that the visual field and the sound must relate. Experience has shown that the width of the screen should be repeated between the edge of the screen and the loudspeaker beside it. This allows a conventional 60-degree pattern in a reasonably sized room. The room should be acoustically symmetrical from left to right, so that sound does not change as it moves from one speaker to another.

For multichannel recording in small studios, a 60-degree angle between speakers is still recommended with a centre speaker, on the centre line (ideally behind a perforated screen), but above if this is the only answer. In four-channel stereo the surround is at the side and back, in 5.1 the two surround speakers are either side at a 110-degree angle from centre. The sub-bass is situated for best response. Dolby Surround software available for small music and audio studios has facilities for calibrating the acoustic monitoring levels within the production suite itself.

For cinema released films, Dolby license out their Dolby Digital and Dolby Stereo systems, which allow control of settings and speaker arrangements to standardize monitoring in the re-recording theatre. To ensure cinemas reproduce this audio to the best possible advantage, exhibitors can sign up to the THX theatre programme, which will offer a certificate and continuous monitoring to ensure their theatres reach high picture and audio quality.

Acoustics and reverberation

Reverberation affects the quality of reproduced sound and so it is a vital consideration in the design of any sound listening room, whether it be a studio or a cinema. Reverberation time is measured by the time taken for a sound pressure to drop by 60 dB; this can vary between 0.2 and 1 second depending on the size and the treatment of a studio. This reverberation time, however, must be the same at all frequencies to produce a good acoustic, and not just correct at some mid-reference point. It is not unusual for a room to have a short reverberation time (RT60) at high frequencies, but a longer one at low frequencies. This will, among other problems, produce an inaccurate stereo image. High-frequency reverberation can be treated with simple surface treatments. However, low-frequency problems often require changes in structural design.

Background noise

It is important that the audio post production studio environment is quiet. Additional noise is distracting and can affect the ability of the sound mixer to hear accurately, although when mixing mono material, it is reasonably easy to ignore background noise as the brain tends to eliminate sounds not originating in the monitoring speaker. With stereo and surround sound, however, background noise becomes more disturbing and it is difficult to ignore the distractions of ambient sound within a room (such as video players with cooling fans, buzzing power transformers, video monitors, air-conditioning and passing traffic).

To assess how distracting external noise can be, architects use noise contour curves (NC curves). Measurements of the acoustic noise of a room are made, using special octave wide filters. The NC rating is weighted to allow higher levels of noise at low frequencies, matching the sensitivity of the ear. A noise level of NC25 is a practical and acceptable level, making it just possible to hear fingers rubbing together with arms at the side. (Figures between NC5 and NC10 are considered very good.) To achieve a suitable figure, noisy equipment should be removed from the monitoring environment.

After an area has been acoustically treated and the furnishings, equipment and speakers mounted, it is usual to take frequency response measurements in order to check the acoustics in the room. If problems do arise, it is possible to electrically equalize the monitoring system, by modifying the frequency response of the speakers, to produce the desired acoustic response. However, this technique will not remedy, for example, deep notches in the frequency response, poor sound diffusion and poor stereo imaging. The electrical equalization of monitoring systems in small studios may not, therefore, necessarily be the answer to acoustic problems. It may well appear to produce an improvement at one point in a small room, but at other positions in the room problems may well be exaggerated. However, these techniques can be used satisfactorily in large environments such as cinemas.

Workstation rooms

Editing rooms containing post production workstations are often merely adapted offices. Unfortunately, many of these rooms have very poor acoustics and insulation, and may be far from ideal for monitoring sound. They may, of course, be used only for checking sound and not mixing; however, many low-budget productions are edited and then mixed in this same room. Unfortunately, these work-rooms tend to be small and have little or no acoustic treatment; this creates undesirable chatter or slap echoes and standing waves, as well as unwanted resonant frequencies, making the sound far from neutral. If soundtracks are mixed in a poor acoustic environment, which is, for example, bass deficient, an editor might well decide to compensate for this apparent problem in the soundtrack by increasing the bass equalization controls. This compensation will soon become apparent once the track is played elsewhere – where it may well sound too 'bassy' and indistinct. It may by now be too late to fix the problem.

Offices and workrooms tend to be rectangular in shape, as this is the most cost-effective way of building, which is not ideal. Some manufacturers of acoustic surfaces offer kits to improve acoustics specifically for post production areas. Each type of wall surface offered will reduce specific problems. Not only can quality be improved, but listener fatigue reduced. Acoustics and speakers must be neutral. Small loudspeakers perched on desks beside workstations are favoured in these situations, giving a 'close-up' sound. For editing, this may give confidence and this can be true sound if it avoids early acoustic reflections from, say, a work surface that may discolour sound. But near-field speakers are not capable of reaching the standards of larger studio monitors; in particular, they can suffer from bass deficiencies because of their small size. The sound field may also be very narrow and slight shifts in head position can produce marked changes in frequency response.

To provide a standard for small room monitoring, the THX organization's 'PM3' certification is available to produce high-quality studio environments.

The importance of listening levels

When mixing any type of programme material, it is important to monitor it at the correct volume. Some organizations label their loudspeakers with a health warning that high levels of sound can damage your hearing. Sound levels should never go continuously above 90 dB or damage will result, and in many countries at these levels ear protectors must be worn!

Monitoring levels should be determined by the level at which the material will eventually be reproduced. There is a strong temptation in the sound studio to monitor at high levels, so as to produce an

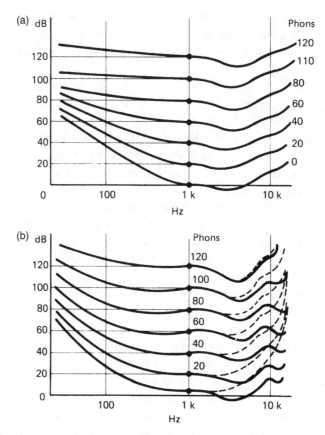

Figure 15.2 Equal loudness curves demonstrating that the ear is more sensitive to changes in volume at middle and high frequencies. (a) Fletcher–Munsen curves. (b) Robinson–Dadson curves include the age of the listener: unbroken line, a 20-year-old; broken line, a 60-year-old.

apparently finer and more impressive sound. However, since the ear's frequency response is not flat, but varies considerably with listening levels, the temptation should be resisted. While very loud monitoring levels tend to assist in discovering technical problems and aid concentration, they will not assist in creating a good sound balance. It can be all too easy to turn up the monitoring levels while being enthusiastic and enjoying a programme, forgetting the problems this will create. Once the sound level is set at the start of a mix, it should never be altered; it should be set to a normal reproducing level.

Listening levels for film audio post production work are set out in the International Standards Institute recommendations and the Society of Motion Picture and Television Engineers' publications, and are widely accepted. Film mixing rooms will try to emulate the motion picture theatre, both from an acoustic point of view and also visually. It is important to see images on a large screen, so as to judge the overall affect of a sound mix. Details in pictures that have synchronous sound can also be viewed with accuracy. A sound pressure level of 85 dB has been generally adopted, which is recommended to be 6 dB below the level at which an optical soundtrack will clip. The introduction of digital soundtracks in films has led to increased reproduction levels in some cinemas, no doubt because of the elimination of system background noise. A lower level of 78 dB is recommended for television monitoring. Although standards do exist in the huge video and television industries, they are less widely adopted in this more budget-conscious sector. Neither do licensing organizations control quality as, for example, the Dolby organization or Digital Theatre Systems do when soundtracks are recorded for film release.

In television, the monitoring environment should try to simulate the listening conditions of the average living room, although many do not. In Europe this size is likely to be about $5 \times 6 \times 2.5$ metres, with the reverberation time having little dependency on frequency (given average furnishings it is about 0.5 seconds). Since the majority of television viewers live in cities, where environmental noise is high and sound insulation poor, there is likely to be a high level of distracting noise. This is countered by turning up the volume, although consideration for the neighbours will restrict this – giving an available sound volume range of about 35 dB in the worst case. An appropriate listening level in the home is likely to fall, at most, between 75 and 78 dB (compared to 85 dB plus in the motion picture theatre).

Television control rooms designed for stereo sound monitoring are likely to have:

● Near-field sound monitoring with viewing by video monitor.
● Speakers placed at three widths apart on either side of the screen.
● Speakers subtending an angle of 60 degrees.
● Facilities to check mixes on high-quality and poor-quality speakers.

Dynamic range, scale in decibels (these figures are very approximate)

	130 dB	Threshold of pain
		Sound level in clubs. Illegal in some countries
	120 dB	'Feeling' sound in the ears
	110–115 dB	Loudest sound on digital film soundtrack
	90 dB	Loudest sound on an analogue film soundtrack
	80 dB	Loud television set
	70 dB	Conversational dialogue
	40–50 dB	City background
	35–40 dB	City flat background
	35 dB	Blimped film camera at 1 metre
	30 dB	Quiet countryside background
	25–30 dB	Quiet background in modern cinema
	25 dB	Fingers rubbing together at arm's length
	0 dB	Threshold of hearing

Dynamic range digital soundtrack spans 120 dB to 35 dB.
Dynamic range television transmissions spans 110–115 dB to 35 dB.

Recording mixers shouldn't attempt to mix at continuously high levels, as permanent damage to hearing may result

Figure 15.3 Relative sound levels.

- A reverberation time of less than 0.5 seconds.
- The operator closer to the loudspeakers than the surrounding wall.
- A listening level set to about 75 dB.

The introduction of widescreen television has increased the effectiveness of stereo sound and introduced the option of multichannel sound systems.

Visual monitoring of recording levels

Loudspeakers in studios provide aural monitoring of the signal. To accurately record these sounds it is necessary to have a precise form of visual metering. These two forms of monitor, the loudspeakers and the meter, must be carefully matched. The average recording level must correspond to a comfortable, recommended, relevant monitoring level within the studio environment. In film re-recording this is carefully controlled by the licensing organization, such as Dolby, who control the process and even provide their own special metering facilities to ensure the recording system does not go into distortion or produces phase errors.

Neither the monitoring volume nor the meter sensitivity should be changed in a system if the sound level is to remain consistent. When mixing, it is usual for sound to be judged almost entirely from the monitoring systems, with only occasional reference to the metering, which provides a means of calibrating the ear. Two types of meters have been classically used for monitoring sound, and much is written about them in other books. They are found both on recording consoles and on the 'meter screens' of workstations.

The VU meter

VU meters are constructed in virtually the same way as AC volt meters and measure the mean average or RMS voltage of a sound signal. It is traditionally seen as a needle meter.

For over 60 years the VU meter has been the American standard for visual monitoring. If the signal is intermittent, such as speech, the VU meter will indicate an average value. This will be considerably lower than the instantaneous maximum levels that are found in the material (to compensate this, engineers ride dialogue perhaps 3–5 dB below the music). The VU meter was never intended to indicate peak distortions or to indicate noise. Its advantages are in its ability to monitor mixed programme material, giving an apparent indication of loudness (which peak meters will not). VU meters often incorporate red LED indicators to indicate when peaks are reached.

The peak programme meter

The peak programme meter (PPM) is not quite as old as the VU meter but its 50-year-old standards are still in use. It is much favoured in Europe, where it comes in varying national standards. It provides a more accurate assessment of overload and transmitter saturation, indicating programme peaks instead

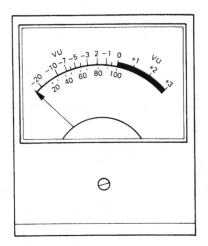

Figure 15.4 A VU meter (volume unit meter), traditional needle display.

of their RMS value. It overcomes the shortcomings of the VU meter by holding peak levels, but it gives no indication of loudness. One form is calibrated between 0 and 7, with a 4 dB difference between each of the equally spaced gradations, but 6 dB between the gradations 0 and 1 and 1 and 2. They all exhibit similar characteristics but with different calibrations.

When peak signals occur they are held in the circuit and other information is ignored. In order to read the meter, it is therefore necessary to wait until the peak hold circuit decays. The peak programme meter is normally lined up to 8 dB below the peak modulation point.

Bar graphs

Bar graph displays use light segments that are illuminated to show level in workstations. These usually display peak readings, although needle-style meters are sometimes recreated. The colour of the display may change with high levels of modulation. Generally, changes in levels and breaks in sound

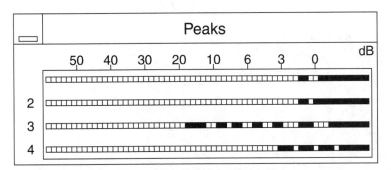

Figure 15.5 Workstation display of a peak reading meter. On overload the display turns red. The H key on the keyboard will display the meter horizontally, the V key vertically.

are easier to detect on this type of meter than on a needle meter. In addition, it is easier to see very low levels of sound on bar meters, as they often show significant indications at 40 dB below peak. In contrast, a PPM cannot show much below 25 dB and the VU much below 12 dB. They are ideal in multi-channel surround set-ups, where many separate meters can be grouped together for easy reading.

Meter displays can be placed within video monitors, but there are a number of drawbacks:

- Visibility is very dependent on the content of the picture.
- To reduce the complexity of the meters, some manufacturers do not include scale markings, and this can make these types of meter difficult to line up.
- Because of the need for burnt-in timecode in audio post production, meters have to be placed at the sides of the screen. With timecode information and meters, TV monitors can become very cluttered.

Mono and stereo metering

Stereo sound recording systems require meters for each channel, with often an additional mono meter to monitor sound for the many monophonic systems that exist. It is normal to make visual meter and aural monitor checks by adding both signals together. In this situation, it might seem reasonable to expect the left and right signals to add up to roughly twice the value of left and right signals together. However, in practice, this increase in level will vary with the degree of correlation between the two signals. If, for example, the two separate stereo signals were out of phase with each other, a combined mono signal would be less than either of the two individual signals. Only when the 'stereo' signals are identical, of equal amplitude, with 100 per cent correlation, such as a panned mono source in the centre of the sound field, equally divided between the two, will the combined signal be double that of the individual ones (an increase of 6 dB).

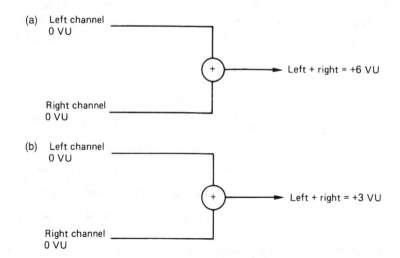

Figure 15.6 (a) Total correlation between left and right. (b) No correlation between left and right.

When there is no correlation between the channels, the sum level of equal-amplitude signals is 3 dB higher than either input channel. Therefore, in a worst case situation, there is a 3 dB difference between mono and stereo audio of equal amplitude. Different organizations take different attitudes to the problem. Much depends on the type of material being recorded (whether it is 'true' two-channel A–B or M/S stereo), the recording techniques and the correlation. In the UK, the BBC uses a 3 dB level drop (attenuation) for a combined mono signal, while NBC makes no attenuation at all. But the figure must be standardized throughout a network. It is designated an M number. In Canada M0 is used, in Europe M3 and M6.

Phase metering

In right and left channel systems, 1 dB of level difference between the left and right channels will be hardly detectable, although a 3 or 4 dB error will cause the stereo image to be slightly one-sided. This condition does not cause a major problem for stereo listeners, but should the discrepancy be caused by phase errors between the two signals, this can produce difficulties for mono listeners. Errors in phase are essentially errors in time between two sources, caused, for example, when an analogue stereo recording is misaligned; in this situation, audio on one track is heard fractionally before audio on the other. The different timing is expressed as an angle. The more complicated and longer the programme chain, the more chance there is of phase problems building up. Problems are most likely to occur in analogue systems or when analogue and digital are mixed in part of the audio chain.

Surround sound meter

The most straightforward and simplest way to display the relationships between signals in surround sound is to use a circular display, which appears as a jellyfish pattern. Three points at the top of the display indicate the front speakers; the remaining sources are the rear sources. Five separate bar meters indicate the level of each channel.

Photographic analogue recording meters

Like digital recording, photographic optical recording systems tend to clip very quickly when oversaturated. To ensure that a photographic Dolby Stereo soundtrack does not go into distortion, a special 'optical clash simulator' is used, simulating the point at which the recording valve ribbons touch. Lights provide an indication of 10-millisecond clash, which is undetectable, and 100-millisecond clash, which is unacceptable.

16 Mixing and processing equipment

Tim Amyes

In the audio post production studio, the 'mixing surface' is the main working area and at the heart of the sound system. This mixing device may be a selection of pages from an audio workstation specifically devoted to sound mixing or it may be a separate 'control surface'. This is variously described by manufacturers as a mixing console, board, desk or even human interface device! Whatever the system, the person who is mixing the sounds should be in a position to hear the best possible sound, to view the picture comfortably, and to have good access to the various facilities on the mixer. There may be other devices needed for additional editing and tracklaying, such as a workstation, CD players, DAT players and sound processing gear – not built into the mixing console, but close at hand. Such equipment is called outboard equipment, equipment within the console being known as inboard. In the large specialist audio studio, where little soundtrack preparation takes place and only mixing, the mixer may not require access to equipment such as the picture players or audio machines, so these are kept in a soundproofed room close to, but away from, the mixing environment.

The mixing console

Mixing facilities, whether in stand-alone consoles or as part of a workstation, are used to hold together various soundtracks, process them and produce a 'final mix'. This mix may be in mono, two-track stereo or surround sound. It may be used in motion picture production, television or even video game productions. The ever increasing demand for multichannel formats has led to an increasing number of channels and tracks needing to be manipulated. Digital sound recording consoles and their equivalent audio workstations have, through automation, made complicated mixing easier. In audio post production, the sound mixer's time is spent in viewing the pictures and combining this with movement of the hand on the mixing console to produce the desired sound – it must be easy to translate 'thought from perception into action'.

Action is most quickly achieved in mixing by physically moving the faders and controls in the traditional manner on a mixing console. However, the same can be achieved within an audio workstation using the sound mixing pages. Here, a mouse or keyboard controls the audio.

Although this may be satisfactory in a music session, where the controls can be set and then left throughout a mix, in audio post production level and equalization controls may need to be altered all the time and at the same time – by at least 10 fingers of one hand (which is where automation comes in)! Unfortunately, mixing using a computer is difficult, a mouse takes up the whole hand, which is the limiting factor – allowing only one fader or parameter to be adjusted with the cursor at any one time. Even so, with time excellent mixes can be produced.

This problem can be solved by making use of a MIDI 'controller mapping' mixing device as an add-on. This provides a stand-alone mixing console of faders laid out in the traditional way, giving 'tactile' control. Conventional mixing can now take place and fader moves are 'mapped' (followed) by the workstation's own graphical faders onscreen. These digital devices also remember and recall the positions of the faders and processors as they are changed through the mix. The changed parameters of additional plug-ins and software facilities are also remembered. Being digital, these maintain the integrity of the digital sound.

Manufacturers who make both workstations and recording consoles offer the most sophisticated integrated devices, but at a high cost – perhaps three times the price of a high-quality broadcast digital recorder.

As an audio post production project progresses, the programme material is mixed to picture and the programme level is controlled to stop overload distortion, which is monitored by meters. The meters need to be viewed easily at a glance and ideally should be in line with the screen or be part of the screen. Similarly, all controls should be close to hand and easy to find, without having to glance down, away from the screen, where action and sound are being related. In designing traditional consoles, manufacturers make the controls neither too small to be difficult to grasp, nor so large that they take up too much space. The length of consoles can be reasonably easily controlled, but it is the back-to-front distance that is a particular problem. This is where the main controls of the console are to be found. Each sound source has its own fader to control the volume and further controls to adapt the incoming sound and send the sound to other sources. From the design engineer's point, consoles using digital sound processing can be more easily configured than analogue consoles. Modern digital signal processing has made these digital consoles cheaper and more versatile than their analogue counterparts. A 40-channel fully automated console can be purchased for little more than the cost of an industrial digital video recorder. In the most versatile digital desks it is possible for end-users to build their own console in software form, almost from scratch. The custom-built console, which became far too expensive in analogue form some years ago, is available again in digital form with all the advantages of digital processing, but at a price!

In all sound consoles each channel will have its own fader and within a console many channels or input modules will be similar. In digital consoles the controls can double up to operate various different functions, selected and shown in detail on a channel's own video display unit. Alternatively, one set of

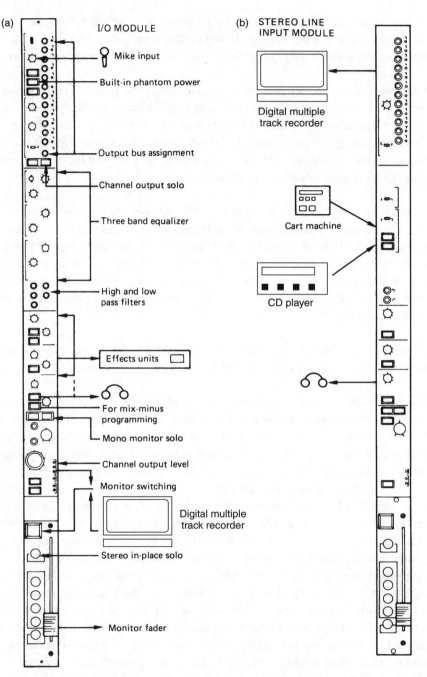

Figure 16.1 (a) An in/out module for a multichannel analogue broadcast desk capable of surround mixing. (b) Input module with limited facilities designed for sources that are already equalized (courtesy of Sony Broadcast & Communications).

functions can be shown in detail on a video display screen, made assignable to the various channels as required. This means only one function can be operated at a time.

The number of input channels required for an audio post production console depends on its purpose. In a small mono or stereo operation, perhaps using a modular eight-track recording system, only 12 channels may be necessary. In a specialist motion picture theatre, where three mixers are sitting at a console, 70 or 80 automated channels may be required.

Types of mixing console

Two types of console are used in audio post production, split consoles and in-line consoles. Split consoles have two sections: one is the simple mixing section sending mixed signals to the recorder, the other part is a separate 'mixer' operating the monitoring facilities. This allows various recorded signals to be routed back and also heard on playback. In addition, this allows some tracks (for example, music and effects) to be monitored but not recorded, while actually recording other parts of a mix, perhaps dialogue. This is called a 'monitor mix'. In-line consoles provide this facility in a different way.

In-line consoles

In-line consoles use what are known as in/out modules.

These contain both channel controls and monitor controls. The audio signal is sent to a specific track on the multichannel recorder, and this track is returned on playback to the same channel. These types of desks are regularly used in more elaborate music recording and in audio post production work, where they can be configured for surround sound recording. In music recording, many separate microphone outputs are sent to many inputs of a recorder; these are returned to the same fader for replay.

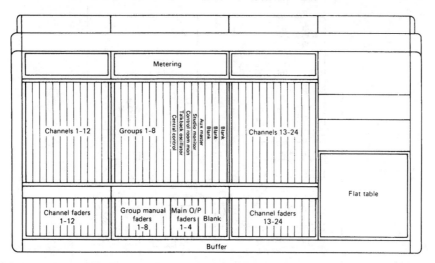

Figure 16.2 A 24-channel analogue in-line desk with four main outputs suitable for surround mixing.

However, this completely flexible switching can be replaced in audio post production by more suitable group routing, where sound is fed to specific master modules. These groups, positioned at the centre of the desk, can then be dedicated to particular functions, such as music, effects and dialogue – and be routed to record or monitor the appropriate tracks as required. These group outputs can accommodate multichannel surround sound.

Inputs

Sound can be delivered to a mixer or workstation at microphone level, at a standard analogue line level or in a digital form. In audio post production there is only a limited requirement for microphone inputs. Most of the material comes from pre-recorded sources, such as hard disk recorders, DAT tapes, CD players or the multiple tracks from an audio workstation. If these pieces of digital equipment are to maintain their highest quality it is preferable that the digital path is followed. All should be synchronized to a master generator (in a small installation, this is likely to be the recording console or workstation), sometimes called the word clock master. This generates a synchronous pulse and allows the other devices connected to it to synchronize data. If any device is not synchronized to the system, drop-out and other noises or glitches may well appear. Well-designed digital mixing systems will offer the necessary options for various digital sound interfaces AES/EBU, TDIF, etc. and sampling rates of 44.1, 48 or 96 kHz or more with quantization of 16, 20 and 24 bits.

In operations, the audio is first passed through the volume control or fader, to adjust the level of sound. In mono, this is simply routed to the desk's output, but in stereo and surround sound, not only can the level be altered, but the sound can also be moved around the sound field. This movement is executed by means of a panoramic potentiometer – each channel having a 'pan' control that moves the sound image between the selected loudspeakers. In its simplest form, the 'panpot' is a rotary control that moves the

Figure 16.3 A joystick control used to move sound around the surround sound field.

image across the two speakers. In a surround mixing system, a joystick (Figure 16.3) or mouse allows the sound to be moved around the multichannel sound field. It is also possible with experience to use a system with three separate rotary panpots. The first pans sound left–centre–right, the second from front to side pan, and the third from left surround to right surround. This facility may be selectable rather than available on every channel. These positions may be automatically memorized within the mixer itself through console automation. Panning in surround when recording for Dolby Stereo surround needs careful thought, as this matrixed system has certain limitations, which are described in the next chapter.

The other facilities needed for audio post production on the input channels, which may or may not be memorized, include the following.

Auxiliaries

Each input module is likely to have various separate outputs (after the fader but prior to equalization) in addition to its normal outputs. These are called auxiliary outputs and are used to feed additional auxiliary equipment, such as reverberation and delay units, as well as sending sound to other sources solely for monitoring purposes, such as with monitor mixes. An auxiliary might be used in post synchronization (the re-recording of poor-quality location dialogue in a studio), where it is necessary for the artiste to hear the original dialogue as a guide track; this can be sent via the auxiliaries (sometimes called foldback) to the artiste on headphones.

Auxiliary returns

Some signals that are sent via the auxiliaries need to be returned to the console after, for example, the addition of echo. These processed sounds can be connected to an input channel or returned via a special auxiliary return, in the form of a stereo input, together with a level control, a pan control and routing selection.

Audition

This is another form of auxiliary but one specifically for monitoring. It includes much of the audio chain – for example, reverberation and equalization.

M/S switches

M/S switches allow M + S matrix stereo format recordings to be processed within the desk itself.

Mutes

Mute switches are cut buttons that silence a channel and are often used to cut out background noise when audio signals are not present.

Phase reversal switches

Phase reversal switches merely reverse the phase of the input connector. They can provide quick confirmation that stereo is present.

Pre-fade listen (PFL)

Pressing the PFL button on a channel allows the channel to be heard before the fader is brought up. This can be used to advantage if tracks have unwanted sounds on them. These noises can be heard on PFL with the fader down and lifted when the problem has passed (AFL is after-fade listening).

Solo

This is similar to PFL but when operated, this mutes all other channels, making it particularly easy to hear the selected audio channel.

Width controls

Width controls allow adjustment of the width of the stereo signal from mono to true stereo, and sometimes can be used to give an apparent impression of greater width to the stereo image by feeding a certain amount of the material, out of phase back, into the channel.

Equalizers

Simple equalization consists of treble and bass controls for changing the frequency response of the sounds. Equalizers are invariably used in audio post production to match sounds. They can also be used to give sound apparent distance or perspective – for example, as a performer walks away from a camera his voice will reduce in bass content. This effect can be recreated by careful use of equalization.

The equalization or 'eq' can be made more sophisticated by offering more frequencies for adjustment and by adding control of the width of the frequency band used. This is known as the Q. The Q frequency can be switchable to various selectable values, or it can be continuously adjustable with the variable centre frequency. In the latter form, the equalizer is called a parametric equalizer.

Filter type	Symbol	Application
Parametric	⊶	Boost or cut can be set for a variable width frequency band centred around a selected frequency.
Notch	⋎	A very high Q (narrow band) filter with a very large (effectively infinite) cut value. The gain control is not used with this filter.
Hi shelf	⊏	Allows gain or cut to be set for all the region *above* the boundary frequency. The Q control is not used with this filter.
Lo shelf	⊐	The Lo shelf is like the Hi shelf, except that the controllable region is *below* the boundary frequency.
Hi pass (Lo cut)	⌐	A filter with a rolloff *below* the selected frequency. The Gain and Q controls are not used with this filter.
Lo pass (High cut)	⌐	Like the Hi pass, except the rolloff is above the selected frequency.

Figure 16.4 Graphic representation of some equalizing curves as used in video displays.

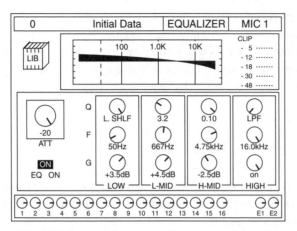

Figure 16.5 An automated digital console: equalization parameters. The equalization function is selected at each channel (courtesy of Yamaha).

There are other forms of 'equalization' found on mixers. These are more specialist functions and may not be available as standard mixing facilities. They come as outboard equipment – as stand-alone boxes (without automation) or computer software plug-ins (with automation).

Notch filters

Notch filters have a high Q and allow a specific frequency to be selected and attenuated. They are used to reduce such problems as camera noise and can be 'tuned' to the appropriate frequency. The offending noise can often be removed without completely destroying the basic sound quality. A depth of at least 20 dB is necessary for successful noise rejection. Tuned to 50/60 Hz rejection, the filter can reduce mains frequency hum, with a further 100/120 Hz notch for its audible second harmonics. A 15 kHz notch filter can be used to reduce the 625-line sync pulse interference noise, radiating from domestic televisions in Europe.

Pass filters

Low-pass filters restrict the high-frequency response of a channel, allowing low frequencies to pass through. They are useful for reducing high-frequency electrical noise from fluorescent lamps and motor systems, as well as high-frequency hiss from recording systems.

High-pass filters restrict the low-frequency spectrum of a system and permit high frequencies to pass. They are particularly useful in reducing low-frequency rumbles such as wind, traffic noise and lighting generator noises.

Graphic equalizers

Graphic equalizers are able to provide equalizing facilities over the entire frequency range, and are often purchased separately from a specialist manufacturer. They use sliding linear faders arranged side by side. This gives a graphical representation of the frequencies selected, hence the name, graphic equalizer. The individual filters are usually fixed and overlap. They give quick and precise visual indications of the equalization used. As with most frequency response filters, graphic equalizers should be provided with a switch to remove the filter in and out of circuit, allowing the operator to check whether the effect being introduced is satisfactory.

Uses of equalizers

Equalization can be used for various reasons:

● To match sounds in discontinuous shooting, where the distance of the microphone to the performers varies as the camera shot changes; with equalization, it is possible to match the dialogue, producing a continuous cohesive sound quality.
● To add clarity to a voice by increasing the mid-range frequencies.
● To correct the deficiencies in a soundtrack if, for example, a voice has been recorded under clothing and is muffled.
● To add aural perspective to a sound that has been recorded without perspective.
● To remove additional unwanted sounds such as wind noise, interference on radio microphones, etc.
● To produce a special effect such as simulating a telephone conversation or a PA system.
● To improve the quality of sound from poor-quality sources such as transmission lines or archive material.
● To improve sound effects by increasing their characteristic frequency – for example, by adding low frequencies to gunshots or punches.
● To 'soften a hard' cut (upcut) by reducing a specific frequency at the start of a clip and then returning it to normal; this is less harsh than a fade.

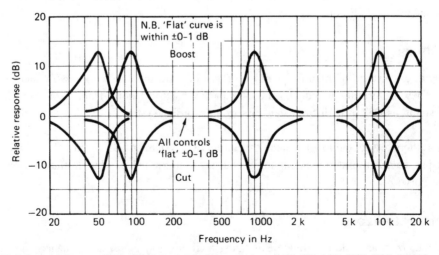

Figure 16.6 A graphic equalizer with selectable frequencies at 50, 90, 160, 300, 500, 900 Hz, 1.6, 3, 5, 9 and 16 kHz (courtesy of Klark-Technik Research).

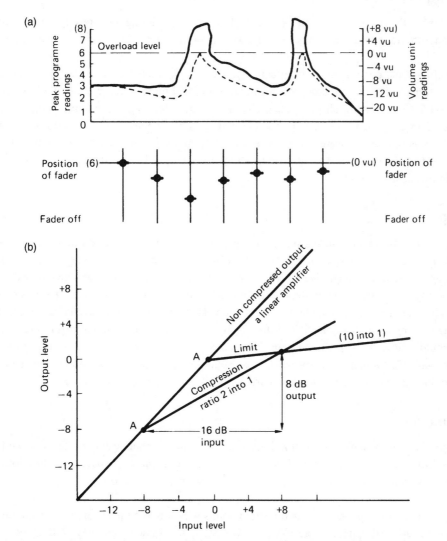

Figure 16.7 (a) Adjustments to a sliding linear fader in manual compression of audio. (b) Sound-level diagram showing the effects of limiting and compression, 'A' being the breakaway threshold point from the linear.

Equalization should, however, be used with care; the ear only too easily becomes accustomed to an unnatural sound 'quality', which it then considers to be the norm. (A poor-quality telephone line may be unpleasant to listen to for the first 5 seconds, but we soon accept it.) It is always sensible to listen to a heavily equalized sound again, without processing, to assess the impact of the particular effect.

Control of dynamics

The range between high and low levels of sound, in reality, is often far greater than is needed for a sound mix. The simplest way to control this is to pull the fader down during loud passages (where overload of the transmitter or the recorder could take place) and to push it up during quiet passages (ensuring that sounds are not lost in the background noise of the system or more likely the reproducing environment).

This manual dynamic control can work effectively, but it is often much simpler and easier to control dynamics automatically, using a compressor/limiter. More and more of these are being built into the channel modules of consoles, allowing each channel separate limiting and compression, often with automation. However, they are usually found, with greater sophistication, as add-on 'outboard equipment' or as plug-in software. They are available from various manufacturers, some who have even returned to using the tube or valve, with its ability to produce a less harsh response to compressed and limited transient sounds.

Limiting

Limiting is used to 'limit' or reduce a signal to a specific level. Further increases in input level results in no further increases in the output. The severity of the reduction is measured as a ratio, usually of 10:1, although some limiters can give ratios of 20:1, 30:1 or even 100:1 (the difference between these ratios is audibly not great). Some limiters may allow fast transients (that is, short fast rising signals) to pass through without being affected by limiting action. This maintains the characteristic of the sound without the limiting effect being audible. Limiting is an invaluable aid in digital recording, where an overload could lead to a distorted and unusable recording. Often, however, limiting is used without thought and valuable audio information can be lost; sounds such as car doors slamming may be limited so as to produce little or no sound at all on a track.

Compression

Compression is a less severe form of limiting, used to produce certain specific effects as well as to control and limit level. The onset of the limiting effect is smooth and progressive. The threshold point is the point at which compression starts, ratios being between 1.5:1 and 10:1.

Low compression ratios operated at low threshold points will preserve the apparent dynamic range of programme material (despite compression). However, at the same time, they will allow a high recording level and thus give a better signal-to-noise ratio. A high ratio at a high threshold point gives similar results, but with the probability of a more noticeable limiting action.

The quality of the compressed sound is very dependent on the speed at which the compressor 'attacks' the incoming sounds – the speed of attack. Slow attack times will result in a softening or easing of the sound. As the attack time lengthens, more high frequencies will pass unattenuated through the system; on speech, this will lead to sibilance. Slow attack times are useful when a considerable amount of compression is needed. When used with a tight ratio, low-frequency sounds will have maximum impact. With deliberate overshoot these sounds will have added punch, which can be useful in recording sound effects. Faster attack times are necessary for speech and can be used to assist in controlling apparent loudness. With very fast release and attack times and high ratios of compression, the low signal content of programme material is raised. This produces a subjective increase in loudness and is particularly useful, if carefully used, in increasing dialogue intelligibility. Unfortunately, if an

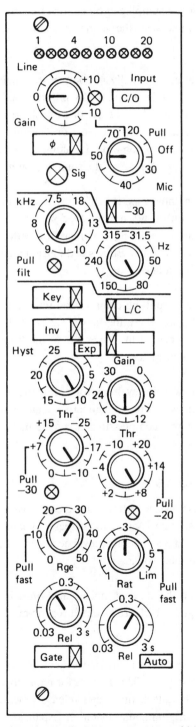

Figure 16.8 A traditional outboard analogue limiter compressor, of high quality, with noise gate and ducker facilities, suitable for rack mounting (courtesy of AMS Neve).

extremely fast release time is used, a pumping or breathing effect becomes apparent as the background level goes up and down in volume with the sound. A traditional compressor limiter is shown in Figure 16.8.

Recovery times can often be set automatically in compressors, being dependent on the level of the input signal. A specific recovery time is automatically programmed when the signal reaches above a certain threshold. As soon as the input falls below this threshold level, the recovery time smoothly changes to a shorter one, perhaps from 10 seconds to 1 or 2 seconds. This is sometimes referred to as the 'gain riding platform'. It is used in some broadcasting transmitters where considerable overall long-term compression is needed, and can affect the sound of carefully mixed programme material.

When compression is used on an entire mix, it is possible to end up with one dominant signal on the track. This can be a particular problem when using loud sound effects or music under dialogue, and may even show itself as pumping or breathing. It is therefore better to compress the various sections of the mix separately rather than all at once. Effects and music under dialogue can even be held down with a compressor rather than with a fader.

Certain compressors are capable of splitting compression into various frequency bands, and this can be particularly useful with dialogue, where the power varies at different frequencies and unnatural effects can be reproduced. These occur at the lower end of the frequency components of speech that form the body of words, and give character.

Normally, compressors incorporate gain level controls, allowing levels to be maintained even when gain reduction takes place. In this way, a direct comparison can be made between the compressed and the uncompressed material. In stereo, compression can create movements in the stereo image, which should be carefully monitored.

Noise gates and expanders

Noise gates and expanders are the inverse of limiters and compressors. A compressor is used to reduce dynamic range, whereas a noise gate is used to increase dynamic range by reducing the quieter passages further. The point where the reduction in level occurs is called the 'gating threshold', and this is adjusted to just above the unwanted sounds.

Normally inserted at the input to the channel, noise gates must be used with care, since low-level sounds such as whispered dialogue, which are essential for the mix, can be treated as unwanted noise and disappear completely!

Some noise gates are available as frequency-selective devices, so that each band of the frequency spectrum can be individually noise-gated. This is ideal for reducing camera noise and other unwanted sounds recorded on location. The attack times and release times of the gate must be as short as possible, to minimize clipping on programme material. Expansion can also be used as a method of noise suppression, by exaggerating the difference between the wanted and unwanted sounds.

Audio restoration devices such as de-clickers are based on noise gates, with the addition of sophisticated equalization, which removes clicks, scratches and thumps from original recordings.

Limiters and compressors are used to:

● Provide overload protection against over-saturating – this is particularly important in digital or optical recording, where over-saturation leads to instant severe distortion.
● Reduce the dynamic range of material to make it more suitable for the final format, (non-hi-fi video cassettes have a very limited dynamic range).
● Automatically reduce the range of sound to a comfortable level for domestic consumption.
● Increase the apparent loudness or create impact.
● Increase the intelligibility of speech and the 'definition' of sound effects.
● In noise gate and expander form, reduce to background noise levels.

Reverberation

Other facilities provided on the console include units designed to add artificial reverberation. Reverberation occurs when sound waves are reflected off surfaces – this is not the same as echo, which is a single reflection of a sound. Reverberation adds colour, character and interest to sound, and sometimes even intelligibility. Reverberation accompanies nearly all the sounds that we hear, and its re-creation can be a very vital part of a sound mix, particularly in speech, when quite often sound that is totally lacking in reverberation seems unnatural. Natural reverberation helps to tell the ears the direction that sound is coming from, the type of environment that the sound was recorded in and approximately how far away it is. It is particularly relevant in stereo recording, but within multichannel formats artificial reverberation must be used with care. Mixes with artificial reverberation can suffer from phasing problems when mixed down to a mono format.

Clapping the hands gives an excellent indication as to the reverberation of an environment, and one often used by sound recordists when they first encounter a new location.

When the hands are clapped, sound radiates in all directions at a rate of about a foot per millisecond. The first sounds to reach the ears come directly from the hands and tell the brain where the sound source is. The next sounds heard come from early reflections, and these will be slightly different from the original sound waves since some of this sound energy will have been absorbed by the surfaces that the sounds have struck (this absorption will depend on frequency). These early reflections can extend from 5 ms up to about 200 ms in a large hall. These reflections will build up quickly to an extremely dense sound, coming from all directions – no longer from just the hands. The ears now receive a slightly different pattern of reflections, at different times, from different directions. The time this effect takes to die away is called the reverberation period. If an electronic device is to recreate reverberation satisfactorily, it must take into account all these parameters.

Essentially, reverberation units take sound and delay it – often through digital delay lines. These delays can range from a few tens of milliseconds to a few seconds. The re-creation of the acoustic of a small room requires only a few milliseconds of delay, whereas a few seconds of delay is required to recreate a public address system. By taking different parts of a digital delay line and applying degrees of feedback and filtering, more sophisticated effects can be created. It is possible to de-tune the delay and, with time slippage, to produce an effect that will split mono sound into pseudo stereo.

The technology used in delay devices also forms the basis for pitch-changing and time-compression devices, where specific sounds can be made to fill predetermined time slots. An example might be where an additional piece of dialogue has to be added to an edited picture, or where a voice-over for a commercial is too long and needs to be shortened. Using a delay device, this sound can be fed into a memory and then read out at a different speed, either faster or slower, but at the same pitch. Reading out data slower than it is fed in means that the system has to repeat itself and there may be an audible hiccup or 'glitch' in sound.

Computerization of mixing operations

The vast complexity of modern recording mixers has led manufacturers to provide computerized aids to mixing, making it possible to memorize every setting at a particular point in a mix as it progresses.

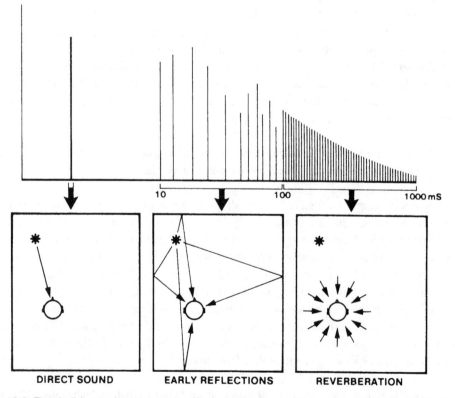

Figure 16.9 The build-up of reverberation from a single sound source (courtesy of Klark-Teknik Research).

Since a sound mix is created against time, it is essential that the mixing console is tied to time. In audio post production, this will be the timecode on the picture master, which could be either SMPTE/EBU or MIDI timecode through an interface. It is important that any automation system is capable of recalling all the necessary information within the time it takes for the workstation or audio machines and the video machine to reach synchronization; 2 MB may be needed to memorize the automation on a 40-input recording console. Automation should never hamper the creativity of the mixer. In the ideal system the operator would need to use no extra controls. Unfortunately this is not completely possible, but it is possible to go a long way towards this goal, much depending on the type of memory storage used and whether it is a reset or a recall system.

Recall merely tells the operator where the controls should be after a successful mix, but it does not reset them to their positions. Recall information is usually in the form of a video display screen, indicating, for example, where equalizers should be placed to correctly reproduce their previous positions. This facility is found in less expensive analogue consoles but not in digital consoles, which can

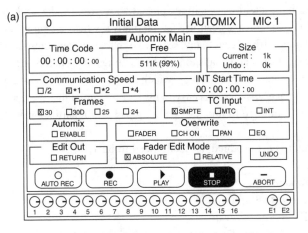

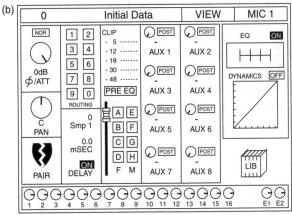

Figure 16.10 An automated digital console: (a) automation display; (b) snapshot of a scene display (courtesy of Yamaha).

offer full reset automation. Digital signal processing can enable all the facilities on a console to be automatically reset – perhaps remembering 6000 different settings.

This total automation system means that sound mixes can be stored in a memory complete without ever having actually been committed to tape or disk. These are called virtual mixes – the recording of the automation, consisting of all the console moves, together with the actual soundtracks. On a workstation these can then be kept for a later date and be replayed together for transfer as a final mix. Reset automation remembers faders' physical positions and puts the motorized controls back to their memorized positions. The smoothness of this movement will depend on the processing power of the console. In less expensive consoles, changes of sound levels may be reset as a series of audible intermittent increments, which is exactly what they are! Much will depend on processing power and cost.

Digital mixers are able to offer complete automation of all their facilities. The ease of routing signals within them means that any functions can be routed anywhere and be remembered.

Traditionally, consoles have additional facilities placed above each fader – physically available on the channel unit or strip. Digital consoles can use an alternative, where an enter button is offered for each facility, giving each channel strip access to one set of assignable master controls. These may be displayed on a video screen giving details and snapshots of mix settings and operating status. Alternatively, each channel may have its own video graphic display, which can be selected to show any of the functions available on the channel. To make such systems easy to use, it is essential that there are not too many separate 'layers of functions' to go through to find the operating page for a particular purpose.

Mixing systems for audio post production ideally need to have:

- All parameters automated.
- Reset automation for virtual mixing.
- Compatible digital audio interfaces.
- Comprehensive automated equalization and compression facilities.
- Integrated transport control and 'punch-in' record facility panel.
- Multichannel sound capabilities.
- Sufficient monitoring for the various 'stems' required.
- Recall routing.
- The ability to store 'snapshots'.
- Data removal compatible with the local systems used.

17 The mix

Tim Amyes

Audio post production is completed with the mixing of the soundtracks. The audio is prepared in the form of separate tracks on a digital workstation. Some of the tracks may have already been mixed and their levels set; much will depend on the time allocated and the experience of the sound editor. In many situations the sound mixer will have also laid and prepared the soundtracks. In productions for theatrical release and television drama, it is not unusual to employ a specialist sound editor or editors, each specializing in a particular area of the sound production. However, the distinction between picture and sound editor and sound editor and mixer is blurring as equipment becomes more versatile, and some mixing consoles incorporate edit functions. It has been argued that with the greater sophistication and efficiency of audio workstations fewer tracks are needed. But with the universal requirement for stereo and the increase in multichannel sound recording (surround sound), this may not happen.

There are both creative and technical aspects to sound mixing. Technically, the most important part of audio post production is to produce a match of the sound to the visuals – so that sound appears to come from the pictures. The visual and aural perspective must be correct; if an artiste moves towards the camera, tradition and realism dictate that the acoustic of his voice should sound as though he is moving towards the camera. This adds interest and variety to the sound (when personal radio microphones are used a decision may be made not to use frequency equalization to change perspective). If an explosion occurs half a mile away, it should sound as though it occurs half a mile away, regardless of how close it was originally recorded. Of course, all rules are there to be broken.

Creatively, audio post production is concerned with mixing the various soundtracks together to produce a cohesive, pleasing, dramatic whole, to the benefit of the pictures and to the director's wishes. A scene, from an audio perspective, is more than just one sound following another (as are the visual pictures), it consists of sounds that knit whole scenes together, adding atmosphere and drama. Sound places pictures in geographical locations and indicates time. The sound of an aircraft passing overhead will join together different visuals to one geographical location, perhaps an airport. Varying levels of the aircraft noise will place the visuals near or far from the airport.

It is important that the audio mixer has respect for both the pictures and the sound, since there is an interaction between the two. The sound image and the vision must be cohesive; if they conflict, the viewer will be lost.

The functions of the mix are:

- To enhance the pictures and point up visual effects.
- To add three-dimensional perspective.
- To help give geographical location to pictures.
- To add dramatic impact.
- To add contrast by changes in volume.
- To produce sound that is intelligible and easy to listen to in every listening environment.

Operation of the controller

During the mix, the soundtracks and the picture are controlled from the workstation or the picture player or a synchronizing controller. Originally, audio post production systems were only capable of running in synchronization from standstill up to speed and into play. With the arrival of magnetic recording, it became possible to replay the sound immediately and to run tracks in interlock backwards and forwards at single and fast speeds. In audio post production this operation is traditionally known as 'rock and roll'. The soundtracks are run forward and mixed, a mistake is made; the system is rolled back, it is then run forward again in synchronization; recording begins again at an appropriate point before the error. To make sure that the insertion into 'record' before the mistake matches the previous mix, a balancing key or send/return switch is used. This compares the sound that is about to be recorded (the send) with the previously mixed sound (the return). If, with the tracks running, the sound that is about to be recorded matches the sound that is already on the track, when the record button is pressed or 'punched in', there will be complete continuity of sound on the final master. This facility needs to be specifically designed as part of a monitoring/mixing system itself and is usually included in fully automated digital consoles. Here, the mix can be 'recorded' as a memory of all the console's switching and fader movements (a virtual mix) rather than as an audio recording. This allows unlimited changes to be made to a sound mix, with the final copy always just one generation away from the original soundtracks.

Mixing, using 'rock and roll' techniques, can limit the continuity of a mix. The system is stopped and started every time a mistake is made; certainly, rehearsals can reduce these problems, but all this takes up time. To help provide clues as to where sounds are to be found during mixing, workstation pages provide cues of all the soundtracks as they pass the timeline and are heard at the monitors.

Console automation

In console automation it is possible for mixes to be memorized in their entirety. Originally designed for the music industry, console automation was confined to the automation of the faders. In audio post production, 'fader-only' automation is not really sufficient. Mixing is a somewhat different process from music recording; the quality of the soundtracks is often altered to match the pictures. This will mean not only changes in level on the channel faders, but also changes in frequency equalization. This can happen at almost every different take when dialogue is shot discontinuously – here, each take in a dialogue sequence may have a different quality (due to varying microphone distance from the speaker). This requires total automation of a console and also requires 'snapshots' of parameters to be taken too. For example, an equalization parameter used for a particular location can be stored and then repeated accurately whenever the location appears again in picture. Routing and patching can also be recalled to repeat the set-up of a particular section or mix.

The virtual mix

In the total recall automated mix, the movement of the faders is memorized in exact synchronization with the soundtrack using the timecode or the MIDI interface of the system. Here, it is unnecessary to actually 'record' the mix as it progresses – all that is necessary is to be able to store and repeat all the actual movements that the mixer makes on the mixing console. These settings are recorded as a data stream; as the fader movements, frequency equalization and other audio processing progresses in synchronization with the picture. Replaying the mix automatically recalls the console's memory and operates all the controls in synchronization with that memory. The original audio is never altered and the operator is free to make updates. He or she merely overrides the movements of the controls where necessary and makes a series of improvements to the mix until the final desired result is achieved. It is no longer necessary to stop a mix when a mistake is made and lose continuity. The mixer now merely returns to the faulty part of the mix and just updates the memory. However, the mix can only remember audio parameters that are interfaced via the console's memory; if audio is fed out of the console and patched through audio outboard equipment, such as compressors, these may not be part of the console's mix system. The movements on the console will be reproduced but no processing will be introduced unless the outboard equipment is tied into the console's operation. Similarly, audio introduced from outside the system will not be reproduced with the mix at a later date unless it is deliberately introduced and synchronized to the system, perhaps by being recorded into the workstation.

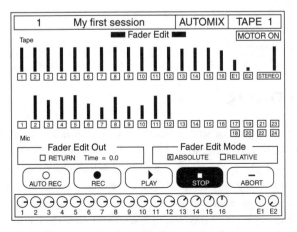

Figure 17.1 Automated digital console: display of fader automation.

When a virtual mix is finally completed, with no possibility of further changes, it will be stored onto tape or disk. Console manufacturers produce a wide range of processing equipment that has automation capability, but the problem of non-automated outboard equipment will always exist. Sound mixing facilities on workstations are able to benefit from a wide range of 'plugs-ins' from well-known manufacturers.

Sound sources for the final mix

In a large audio post production mix, such as for a theatrical film, hundreds of individual tracks may be used to make up the final soundtrack. In a small video production there may only be two tracks available, recorded on an industrial video recorder.

Cue sheets

Track sheets show the various available soundtracks that will make up a mix in graphical form. Once pieces of paper, today screens on modern workstation systems provide all of the information needed to mix tracks. Most valuable is the timeline, which the laid tracks pass across. The tracks may themselves have information written on them and often show a graphical display of the audio waveform.

The process of the mix is as follows:

- View to become familiar with the material in one run without stopping, with the most important soundtracks being played – watch and listen to get a feel of the material before starting to mix.
- Listen to the dialogue tracks to understand the material.
- Decide what needs to be premixed, if anything.
- Premix the dialogue – equalization, noise gating, compression, etc.
- Monitor the other tracks while recording premixes.
- Premix the effects – equalization, compression, etc.
- Premix the music.
- Final mix.
- Review and check the above.

Mixing using DAWs

Mixing soundtracks to produce a final soundtrack within the confines of a DAW (picture and sound or sound only) using only the screen and a mouse to manipulate the audio can be difficult, although much depends on the complexity of the mix. Using a mouse will allow only one track parameter to be adjusted at a time. In some budget workstations there are other problems; it may only be possible to add overall equalization to a complete soundtrack and then be able to move the track, for if it is subsequently moved the equalization may stay where it was! With limited facilities, perhaps using only a four-track stereo system, it is often simpler, when laying tracks into the workstation, to include actual fades and equalization as the tracks are being laid, leaving anything that might be uncertain to the mix. However, it is sensible to divide the tracks into the three traditional distinct groups of dialogue, sound effects and music, which will allow reasonable control over the sounds in the mix. Even within a simple system it is possible to produce a quality mix, particularly if the tracks are carefully prepared.

If the workstation is tied via MIDI to a simple automated digital audio console, the system can work very successfully and the mix will stay as a virtual mix within the workstation, ready to be updated as required or transferred out to the video recorder. An alternative is to use a simple analogue mixer not tied to the workstation and bring the tracks out of the workstation and mix them directly onto a separate recorder or back into the workstation, but this time the mix will be fixed and complete.

In larger systems, where there are a number of tracks being mixed together, they should be subgrouped into various smaller premixes for easier handling. Using a totally automated console or workstation, these can be held in the form of virtual premixes of various tracks. In these systems the mix is memorized as a virtual mix locked by timecode to the tracks. The actual tracks, existing unaltered as different mixes, are tried and recalled. These premixes may well be kept as separate physical recordings for safety purposes. Automated systems do crash!

If a premix is produced without using automation, it will be recorded onto a track or multiple tracks in lock with the complete system. It will then be replayed with other tracks for a further final mix to be made. Unfortunately, the premix is now set and it will now not be possible to alter it, since the levels have already been decided and set. If the mix had been 'recorded' as a virtual mix further changes could have been made. Much will depend on the sophistication of the mixing console as to how many tracks can be

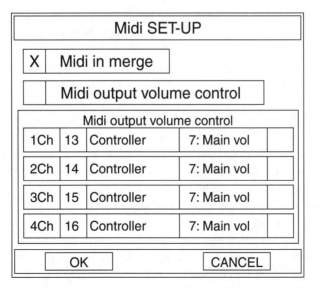

Figure 17.2 A four-channel MIDI display screen for a MIDI workstation interface. The output volume control is selected. If the controller messages are recorded into a MIDI sequencer with the workstation, the entire mix and equalization parameters can be automated.

memorized via a virtual mix. In a complicated theatrically released film, many tracks and their various parameters will need to be remembered; this requires an expensive console with much processing power.

Premixes can be divided down into groups such as dialogue, effects and music, or even further, into premixes of spot effects, foley effects, atmosphere effects, etc. Premixes must be carefully handled, since it is possible for certain sounds to be obscured when the various premixes are finally mixed together. Depending on the equipment, it may be possible to monitor other tracks while the premix takes place. Stereo positioning will also be considered (separate premixing of dialogue, effects and music will allow a foreign language version to be made using the Music and Effects tracks).

In many audio post productions, the dialogue will be the most important part of the mix. A dialogue premix is therefore often recorded first. This allows the mixer to produce the best possible dialogue quality, concentrating specifically on matching sound levels, frequency equalization and stereo positioning. Dialogue is rarely recorded in stereo on location (sometimes in Europe, rarely in the USA). The premix will include panning information for stereo and possibly reverberation, but as many options as possible should be left open for the final mix if the mix cannot be altered later.

When premixing for television, it is often sensible to occasionally reduce the studio monitoring level to confirm that the premixed sounds will not be lost when added to the final mix. This can be a particular problem when heavy effects are being mixed under what will eventually be dialogue. Equal loudness curves show that we perceive sounds differently at different levels, so any premix must be treated with care – only if the mix is held as a virtual premix can it be easily altered. Theatrically recorded mixes are recorded and played (in cinemas) in environments that are carefully calibrated to be at the same volume.

Once the premixes are completed the final mixing begins. Premixes are, perhaps, not so necessary if an automated console is used, although for ease of handling in large projects premixes are still useful.

Using an automation system, the mix for a documentary television programme on an airport might progress as outlined below.

Five or six various effects need to be mixed, plus dialogue tracks. In the first pass, the mixer feels that the background atmosphere mix is correct but the spot effects of the public address system are too loud, and that the plane passing overhead reaches its loudest point a little too late; perhaps the director doesn't agree with him. On the second pass, an attempt is made to improve the public address effect further, but this time the dialogue is slightly obscured. On the third pass, a slight attempt is made to increase the level and improve the intelligibility of the dialogue by reducing the background effects. This, the director likes. Now, all three attempts can be reviewed and compared by memory recall. It is decided that the final mix should consist of the atmos mix from the first pass and the dialogue mix from the second. Now the sound mixer merely selects the automation related to the appropriate fader and parameter movements from the computer, and the final mix is produced.

As the final mix is memorized, any last minute updates of the balance of the tracks can also be carried out.

A second approach is to use sections from the various mixes. The first 30 seconds from mix 1 is used, the next 3 minutes from mix 3 and the rest from mix 2, making the combined final track.

A third approach is to use the computer for virtual premixing. Various soundtracks are taken and mixed as a premix, and then further tracks are added, so building up a final track.

If a non-automated mixing console is used, the mix will progress more carefully; each scene or section will be mixed and agreed upon before the next portion of the soundtrack is mixed. If an analogue mixer is used, care will need to be taken to ensure that premixes are recorded at the correct level to ensure that there is a minimum amount of system noise introduced. The premixes are then mixed and copied onto the final master to produce a final master mix. In a virtual mix, the final soundtrack may be held only as digital data of the varying console parameters and source audio. If this is held on a computer disk rather than magnetic tape there is instant access to any part of the mix.

As the mix progresses it may become necessary to change the position of a soundtrack. Perhaps a piece of voice-over interferes with a sound effect, or perhaps the position of a sound effect is wrong in relation to the visuals. Here it is important to ensure that any processing, such as equalization, is locked to the sound move as well.

The final sound balance of a production needs to conform to the local technical requirements, having:

- The correct relative sound levels.
- The correct dynamic range for reproduction.
- A consistent tonal quality.
- High intelligibility.
- The required perspective.
- The required acoustics.

The final sound balance is the result of mixing together the various discrete elements that make up a soundtrack. Mixing together the final premixes produces the final mix stems. The stems are the various strands that constitute the mix. In a large production these can be many tracks; the surround mix will be held as discrete, screen left, centre and right, together with surround tracks, rather than encoded into two tracks (if it is Dolby Stereo). Other stems will be held as separate dialogue, effects and music tracks.

Keeping the final mix as separate tracks makes it easy to prepare other audio versions of the production, such as airline versions and foreign language versions (using the music and effects tracks and new dialogue).

Mixing in surround

A well-mixed stereo soundtrack can add greatly to the enjoyment of a production, and traditionally most broadcast programmes are produced this way. In stereo the sound is 'captured' within the screen;

with surround sound the audio can escape! Now three-dimensional sound is offered, but with flat two-dimensional pictures.

In music recordings the listener can be positioned in the centre of the action – perhaps even in the orchestra. However, most conventional surround sound music recordings place the listener in front of the musicians. This is a tradition – we expect to be part of the 'viewing and listening' audience.

Surround sound for pictures has similar traditions. The pictures we view are at the front; if we hear dialogue from behind us, we will involuntarily move our heads to the rear to see – the illusion is lost. A jet plane suddenly roaring overhead and instantly appearing onscreen will emphasize the effect both aurally and visually, while a very slow prop-driven aircraft meandering over from behind the audience to the screen will be almost unnoticed, but between these two extremes care must be taken. It is only too easy for surround sound to break the visual illusion created. The vast majority of the information we act upon comes from our visual senses.

It is also disturbing if, for example, we lay loud atmosphere effects in the surround speakers when there is a change of location – this can emphasize a picture cut unnecessarily, which even in mono can be distracting; better to slowly introduce interesting effects. Background sound effects with the occasional heavy noise – for example, building sites – must also be carefully monitored. A smooth background with a sudden loud noise from side or back will again break the illusion of 'reality'.

Film soundtracks are now most commonly mixed in discrete 5.1 surround. However, it is still current practice to produce a Dolby Stereo Surround mix. Here the screen left, screen centre, screen right and surround channels are matrixed into the two stereo channels, called Left total (Lt) and Right total (Rt).

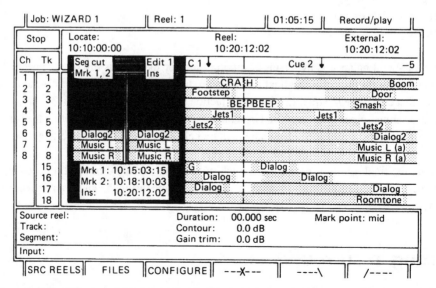

Figure 17.3 A workstation display of the track sheet of a documentary programme. To create an M&E, the mix would not include the three dialogue tracks, but would include the room tone.

This encoded system can suffer from track-to-track crosstalk and phase problems, which sometimes makes compromises necessary in the mix. The surround channel has a reduced response of 100 to 7 kHz, with an adjustable delay of 20 milliseconds. Dolby Stereo Variable Area (SVA) is recorded on photographic film, and has a frequency response up to 12.5 kHz and a slightly limited dynamic range compared to modern digital recording.

Stereo to encoded surround

Many aspects of mixing in surround are the same as for stereo. The encoded tracks are recorded as a stereo pair. This is ideal for broadcasting; without decoding the matrixed audio delivers just stereo sound to the listener, with a decoder full surround. In the mixing room surround demands more equipment; extra amplifiers and speakers are needed and the mixing desk must have at least three and ideally four outputs to feed the encoder. Although the final soundtrack will be recorded on two tracks, when premixing it may be useful to record on four discrete tracks.

A console with only a stereo bus and an auxiliary can be used to create surround sound, but the opportunity to pan the sound within the surround field is limited. The stereo bus can be used for the front channel panning and the auxiliary for the surround. The audio can also be fed to more than one fader, with each routed and 'panning' to a different output.

When mixing with the matrix system, the centre speaker information will be (because of the system design) unavoidably recorded in the right and left decoded channels. Sound panned left or right may appear on the surround channel, and panning between the front and the surround speakers does not always give the expected image movement. Since the quality of the front channels is not the same as the surround channel, there may be noticeable quality changes as the sound is moved around. Similarly, sounds cannot be moved cleanly from the rear speakers to the centre front speaker; in this case, the sound will expand to the left and right speakers and then close into the centre. Neither is it possible to have the same audio signal on both centre and surround channels at the same time, because the left and right total tracks cannot be in phase and out of phase at the same time. Artificial phase reverberation must be carefully controlled too; spectacular effects in surround sound can easily become unintelligible when matrixed.

In surround sound:

- Dialogue should stay in the centre, occasionally moving left or right to follow dramatic action but rarely into the surround. If dialogue does come from one side of the screen, the position of the image will be dependent on where the listener is sitting in relation to the screen. Only for specific effects should the dialogue move.
- Effects can be mixed successfully in surround to give a sense of realism, but care is needed, as irrelevant noises will be distracting.
- The amount of surround will determine how far back the listener is in relation to the screen audio – with more surround the listener appears to go further back.
- Mono effects can be successfully used if divided between centre and surround channels – the effect will appear to come from all four channels (a 2–4 punch). Similarly, by adding stereo reverb left

and right, with part of the dry signal to centre and surrround, there will appear to be an increased width to mono images.

- Excessive, literal panning may well unnecessarily emphasize picture cuts.
- Music should primarily be in the front speakers, with an element in the surround.
- The key to good surround sound is subtlety – surround effects should complement not distract.

When mixing in encoded surround:

- The centre speaker tends to reduce the width to produce a slightly different sound from the two speakers alone, which can be compensated for by making the image slightly wider.
- Some library material may create unexpected sounds from the surround channel. This material may have been artificially spread by the use of out-of-phase signals – this can confuse the decoder.
- Stereo synthezisers can also create problems, particularly if used with dialogue, making mono dialogue appear to come from all speakers at once.

Dolby Surround is the domestic version of the Dolby SVA optical sound system. Found in two-channel broadcast systems and videos, over 40 million domestic decoders have been sold worldwide. It is available to audio post production suites as software interfacing with specific audio workstations for non-theatrical use. Stand-alone units include the Dolby SEU4 encoder and SDU4 decoder.

When mixing with surround for broadcast/domestic use:

- Monitor everything through the surround sound decoder (stereo straight through) in case viewers have left their decoder on.
- Ignore the centre speaker (essential for film), making a more compatible stereo to surround mix.
- Do not clutter the sound field, which will confuse the decoder with blurred images at home and may make dialogue unintelligible.
- For narration, adding a little signal to the surround channel will bring the voice forward.
- Check mono and stereo compatibility – stereo compatibility is not usually a problem; in mono any surround signal will disappear, possibly removing valuable information.

Discrete surround sound systems (not matrixed)

In the late 1980s, an agreement was made in the USA that the minimum number of channels for a digital surround sound system, as opposed to analogue surround sound, should be five, plus a low-frequency omnidirectional channel. This is designated 5.1. Formats are available with up to 10 channels.

These digital surround systems, such as Dolby Digital and DTS, use discrete channels where each soundtrack is recorded in full fidelity. There are fewer limitations to mixing and recording.

The monitor layout is similar, but with at least two surround speakers, as well as the left, centre and right screen speakers. In addition, there is one LFE (low-frequency effect) channel with a limited bandwidth up to 120 Hz, the '.1' of 5.1.

Digital surround systems

- These have more sources, each individually controlled, without compromise – this allows full use of all the individual sources.
- However, this needs extensive monitoring and recording systems.
- They use discrete individual recording tracks, which do not have the frequency limitations of the encoded surround systems.
- They have a better dynamic range than the SVA systems.
- They do not have operational crosstalk or phase problems.
- Compatibility with domestic digital decoders must be considered, as these tend to use some compression, which can affect the quality of a mix.
- Home management systems can lose bass quality by ignoring the '.1' content material recorded only in the LFE, which may be lost in decoding.
- Care still needs to be taken in using the surround sound, as with any surround system. It is now possible, for example, with a discrete system to slowly pan from front to rear, but to the listener this source may appear to be coming at 45 degrees in front not 90 degrees – as the source pans back it may break into two distinct sources heard at slightly different quality. This is because the ears' response to sound from behind is different from its front response.

Compatible mixes for television

Reducing a surround soundtrack to a stereo track creates few technical compatibility problems. Folding down to mono can create problems if essential material has been recorded into the surrounds, as it will be lost. There are, however, other considerations in maintaining compatibility.

Films mixed for cinema exhibition can make full use of all the dramatic power that sound can bring; there are no distractions for the audience, little background noise and no neighbours to complain about the volume. Developments in digital sound recording and reproduction have allowed a final soundtrack to have a very wide dynamic range. The dynamic range recorded may vary from the sound of a gunshot to the silence of a grave. The final mix will be suitable for the environment in which it is to be reproduced – the cinema – but it may not be entirely compatible for stereo and mono reproduction at home.

More and more motion picture feature films, mixed for the cinema, are being distributed for home use; indeed, the majority of the income from feature films is derived from video sales. This means that it is in the home environment, and probably from the speakers of a television, that most soundtracks are likely to be heard. In general, soundtracks mixed for the cinema will, when replayed at home, have the music and effects at too high a level and the dialogue can become inaudible or unintelligible. In addition, the dialogue equalization used in cinema soundtracks to improve articulation and intelligibility can, in the domestic environment, be reproduced with sibilance, giving the impression of distortion.

In the home, the speakers are in a near-field environment, which is flat and narrow with little reverberation. This is incompatible with the motion picture mixing theatre, which offers a widespread image with more reverberation and is a far-field recording environment. As such, the monitoring system is not compatible with stereo television; the large distance between the two screen speakers means also that there is, in effect, 'a hole in the middle' – which does not exist in the near-field environment. Any sounds of long duration tend to sound louder in a film theatre than they really are. Therefore, sounds that are powerful in a loud scene will sound somewhat less at home. Most viewers adjust their television sets to reproduce dialogue at an acceptable level, and with television mixing this is the starting point of the mix. Cinema mixing allows a much greater freedom, with the opportunity to almost lose dialogue under sound effects. Dynamic cinema mixes may therefore not be suitable for television transmission.

Layback

In the re-recording process, the soundtrack is physically separate from the picture. Later, for motion picture theatrical release, it will be finally 'married' to the picture in the form of a multichannel photographic analogue or digital sound print. The soundtrack will be delivered to the audio transfer house most probably on an eight-track multiple DTRS format (Tascam DA88), which will specify the track layout, such as:

Track 1 Left
Track 2 Centre
Track 3 Right
Track 4 Left Surround
Track 5 Right Surround
Track 6 Subwoofer
Track 7 Analogue Dolby Left (matrixed to include centre and surround)
Track 8 Analogue Dolby Right (matrixed to include centre and surround)

It is important to contact the transfer facility to confirm the format and the track layout they will accept before material is delivered to them.

In video production, at some point the audio master soundtrack may well need to be 'laid back' onto the video master for transmission. This might be unnecessary, particularly for news and sport, where the material has been edited into a server. Similarly, the programme may have been mixed and recorded straight back onto the original edited master video for immediate transmission.

Often, however, the soundtrack is held in an audio workstation in the audio post production area with a non-transmission quality picture. The sound now has to be 'married' to the master video picture. The synchronization of the layback is timecode controlled. Since both picture and master soundtrack have the same code, a simple transfer of audio, picture and timecode is all that is needed in chase mode; this could be from a virtual mix held in the mixing desk or the workstation. Multichannel sound can add further problems to the layback. Most video recorders only support four audio tracks; this is not a problem with stereo or matrixed Dolby Stereo, where only two tracks are needed. However, discrete multiple channel surround systems require six separate sound channels (5.1). To enable the necessary six tracks to be recorded on a video recorder, the Dolby Company has developed the Dolby E system. This allows up to eight tracks to be recorded onto the two tracks of a digital video recorder. An audio compression system is used; the encoding system takes one frame to process the sound. This requires a delay to be added to the video picture on replay of one frame. The remaining two audio tracks of the videotape are still available for other programme material.

Music and Effects mixes

Much material that is audio post produced needs to be revoiced into other languages for sales to foreign countries. For this, a mix is created that consists of music and effects only: this is usually referred to as an M&E mix (Music and Effects). It is sold as a 'foreign version' to countries that wish to dub their own language onto the material. The M&E mix consists of all music and effects tracks, as well as vocal effects tracks such as screams, whistles and crowd reaction tracks – in fact, sounds which cannot be identified as specific to a particular country. The balance of the mix should be as faithful a recreation of the original language version as is economically possible. To assist in making a music and effects track, the final mix is used, itself divided into and recorded as three separate tracks of dialogue, music and effects.

In documentary work, a mix will often contain elements of both sync dialogue and voice-over. In this case, the M&E will be made including all production sound, whilst omitting the voice-over.

This voice-over will then be re-read in the appropriate foreign language, with any sync dialogue being subtitled or 'voiced over'. This saves the expense of post synchronizing, which is not a practical proposition for documentary work. Indeed, more and more foreign productions are now being sold with subtitles rather than post synchronized dialogue. However, subtitles can often only summarize dialogue, particularly in fast wordy sequences. This means that the subtleties of the spoken word may be lost.

Creating an M&E is an additional expense that requires more sound mixing time – often, this cannot be justified in television production unless the material is pre-sold to overseas countries (it is a standard requirement in film production). However, if a virtual mix is held of the project, recreating an M&E is not time-consuming, providing the tracks have been well laid.

If M&Es are to be delivered separately, for television, it is likely that the DTRS format (DA88) will be used to provide the stems. A likely track configuration is:

Tracks 1 and 2   Separate fully stereo balanced dialogue
Tracks 3 and 4   Separate stereo music
Tracks 5 and 6   Separate stereo effects
Tracks 7 and 8   Separate narration tracks

Delivery requirements

For higher budget productions such as TV drama and films destined for theatrical release, the M&E is only one element of the *delivery requirements* or *deliverables* as they are often known. Once the final mix is completed, a number of versions of the mix will be made, which the producers will be contractually obliged to supply with the programme master. These versions will primarily be used for overseas sales and archiving purposes. The delivery requirements will specify the type of *stock* to be used for each item: usually formats such as DAT, DA88 and even 35-mm mag stock are used for deliverables, as these have a reasonable degree of universality.

The delivery requirements typically required for a feature mixed in 5.1 are listed below:

- The 5.1 mix master encoded in Dolby SRD.
- The mix encoded in Dolby Stereo (SVA), from which an optical two-track master is made.
- An M&E in 5.1.
- An M&E (LCRS) with 2 dialogue tracks (i) with optional dialogue fx, and (ii) with fold down of dialogue stem as a mono guide.
- A TV mix and TV M&E.
- A stereo DME fold down (dialogue, music and effects).
- A 5.1 master of all music cues/source music (without fader moves).
- A stereo transfer of all music cues/source music used.

18 The transmission and reproduction of audio post production material

Tim Amyes

Once audio post production has been completed, the mixed soundtrack is ready to be distributed. The programme material could be transmitted or reproduced on any format of any quality, and may or may not be in a multichannel form.

Low-quality formats include:

● Video cassette recorders (not hi-fi).
● Websites.
● Computer video games (at 8-bit quantization).

High-quality formats include:

● Video cassettes with hi-fi sound systems.
● Television transmission using digital sound.
● Digital film soundtracks.
● Digital versatile discs.

Between these two quality extremes are:

● Thirty-five-millimetre optical soundtracks with noise reduction.

- Analogue sound television transmission systems.
- Broadband Internet websites (varying quality).

In the cinema, the sound reproducing chain from final mix to released motion picture film is comparatively short and can be of the highest quality. In television, the chain is longer.

The cinema chain

In the motion picture theatre, the sound system consists of two distinct chains, the A and B chains. The A chain is the film with its soundtrack and the equipment to reproduce it (such as is licensed by the Dolby company); the B chain consists of the remaining path the sound follows from amplifiers to loudspeakers and includes the cinema acoustics, which affect the sound quality. Great efforts are made to ensure the sound emerges in the cinema as it was heard at the dubbing stage.

The cinema can be one of the best environments in which to hear recorded sound. Many of the specially built multiplexes of today have excellent sound reproducing equipment and superb acoustics. Modern projectors and their digital sound systems are capable of exceptional quality. To encourage professional presentations one company, THX, tests and certifies cinemas that reach a high standard of projection and audio quality (the B chain). The aim is to ensure that the cinema auditorium reproduces a soundtrack the way the filmmaker intended. For this, the theatre must have a low noise level below NC30dB, be well isolated from adjacent auditoriums and other noises, and have an acceptable reverberation characteristic (particularly without annoying 'slap echoes'). Visually, the viewing angle from the furthest seat in the auditorium must be no greater than 36 degrees. Screen image distortion and illumination is also checked. Both film and video (d-Cinema) projection systems are checked.

The images produced by the best modern video projectors today are almost indistinguishable in quality from the images thrown by a film projector. Images must have high definition, excellent contrast, good colour rendering and be bright; to produce all this from a video projector requires state-of-the-art technology (which simple mechanical film projectors do not need!). Video images often suffer from insufficient illumination; lamp technology is available to create the necessary lumens, but the heat the lamp produces is excessive. A single 35-mm projection frame stays in front of the lamp for just 0.041 seconds. In the video projector the device producing the picture is held in front of the lamp permanently. Digital Micromirror Devices (DMD) provide a solution to heat problems. Light is not passed through the device, as with a liquid crystal display (LCD) device, but is reflected through a multitude of small, coloured metal mirrors, which in turn reflect the heat when light is reflected onto the screen. Lamp life is unfortunately short, and the projectors cost about five times more than a comparable film projector.

It is not surprising that exhibitors are resisting the change to d-Cinema. For distributors, however, there are considerable advantages. Tape duplication is cheap and easier to control. It will no longer be necessary to provide expensive prints, which have a limited life span and are also heavy and costly to transport.

The d-Cinema system is entirely digital, marking the end of, amongst other things, analogue sound. If distribution is via satellite rather than, as at present, by using tape, sports, news, concerts and live events could be regularly relayed to cinemas – as indeed regularly happened in some cinemas when television was in its infancy over half a century ago!

Digital television

In the early 1980s, much improved television pictures and sound were technically possible through improved analogue transmission systems; however, the greater bandwidth needed was just not available. Eventually, researchers offered ways of transmitting high definition television within existing analogue TV bandwidths using digital techniques (DTV).

In the USA, the Federal Communications Commission (FCC) asked those involved in the research to produce a single standard for digital TV broadcasting. The standards proposed have been prepared by the Advanced Television Systems Committee (ATSC), and HDTV is now being transmitted. In Europe, HDTV is still only an originating format, which cannot yet be broadcast. But most countries have or will soon have new standards for HDTV digital broadcasting. The new generation of television receivers produced will not only receive DTV broadcasts through the air, but also via satellite and cable. It seems that there will not be an international standard for transmission, but several standards.

Television transmission

Replayed from the server or video recorder, audio post produced sound and edited pictures for broadcasting are routed to the transmission area of the television complex, where they are monitored, equalized and then fed to their correct destination (usually at an audio sampling rate of 32 kHz). This area also routes tie lines, intercoms and control signals, and provides all the necessary equipment needed to generate pulses for scanning, coding and synchronizing video and audio throughout the site.

Sound in stereo form is usually sent in a simple left/right format, with or without surround matrixed signals, allowing anyone with the correct equipment to pick up the surround sound. Care has to be taken as, unfortunately, matrix signals suffer loss of stereo separation rather more easily than discrete left and right signals.

Any television organization using stereo must produce a decibel attenuation standard for the mono sum programme signals (produced from the left plus right stereo signals), and this figure must be standardized throughout the transmitting network. It is designated an M number. In North America M0 is used, in the BBC and in Europe M3. The problems of routing stereo signals are very much greater than routing mono; not only must the signals maintain the same level left and right, but they must also maintain phase. This is particularly important if the transmitting chain is long and complicated. In America, for example, a large network broadcasting company may have two network distribution centres, one on the east coast and one on the west coast, each distributing sound and pictures over satellite and terrestrial systems. When the links are made, the effects of individual degradations are cumulative and the tasks of discovering the source of a problem can be enormous. The introduction of digital distribution has reduced these problems dramatically, with the added advantages of reduced maintenance and adjustment costs.

The video signal leaves the studio centre by a microwave link or a coaxial line. The accompanied sound is carried on a high-quality digital AES/EBU audio line or within the video signal using a digital technique, which inserts the audio into the line-synchronizing pulses (sound in syncs). This has four major advantages:

- The quality is good, since it is digital.
- It does not require a separate sound circuit and is therefore cheaper.
- There is no possibility of the wrong picture source going with the wrong sound source.
- Picture and sound stay in synchronization even if a delay is added to a source to synchronize with another.

Television chain – transmission

Television signals can be transmitted into the home through cables, both broadband and telephone (the Internet), through the atmosphere from earthbound digital or analogue terrestrial transmitters, or through space, from satellites.

The signals transmitted through the atmosphere or space are essentially radiated frequencies vibrating above the frequency of the audio spectrum, from 20 kHz to 13 MHz and beyond. For terrestrial analogue television transmission, the video signals are frequency modulated, rather like the way video signals are recorded. The sound is transmitted on a slightly different frequency to that of the vision, allowing sound and vision to be processed separately. Audio signals up to 15 kHz can be transmitted.

Television transmissions have to contend with absorption; waves of low frequency are readily absorbed by the earth and objects in their path. Higher frequencies are not, and are used in broadcasting, taking advantage of the ionosphere (layers of ionized air between 100 and 1000 km above the earth). These layers are used to bounce radio waves back to earth – rather like a radio mirror. However, these reflections back to earth begin to fail at about 100 MHz, and completely disappear at about 2 GHz. At this point, the waves travel straight through the ionosphere. Satellites use these high frequencies to receive their signals from ground stations via microwave links. Since these signals project straight through the ionosphere (in the space above the ionosphere there are no transmitting losses), transmission can be over very long distances.

Arriving at the satellite, the signal is converted to the satellite's transmitting frequency and then sent back down to earth. It was the introduction of compression techniques such as MPEG that made satellite transmissions cost-effective. A typical satellite might be able to transmit 18 different analogue channels, but with digital processing, for each analogue channel 30-plus individual 312-line VHS quality channels can be transmitted.

A satellite can cover an area that would normally require several hundred terrestrial transmitters. Whereas over 600 stations might be necessary to cover the UK using earth-bound terrestrial television transmissions, one satellite can do the same job, with its footprint covering much of Europe as well, although there are obviously language problems and cultural barriers.

Multiple audio channels in the home

Stereo sound for television can be transmitted in various ways, but the main concerns of any system must be to ensure that the stereo sound does not affect the quality of the mono transmission, and that the transmitted area in stereo is the same as for mono. In the UK, the digital audio signal is quantized at 14 bits and then companded to only 10 bits/second. The system used is called NICAM (Near Instantaneous Compansion and Audio Multiplex). Two-channel Dolby matrixed stereo programmes can be transmitted in stereo format to produce four-channel surround sound at home; these are reproduced on four separate speakers.

Latest in the Dolby armoury is Dolby Virtual Speaker Technology, which attempts to provide true-to-life surround sound from two speakers. Using crosstalk cancellation and frequency modification, Dolby provide an illusion of surround sound in small environments from a two-channel stereo image.

Dolby Surround Sound

Dolby Surround Sound is the consumer side of Dolby SVA, which is decoded domestically. It is available in consumer equipment in both simple passive and active versions. The cheaper passive version, without amplifying circuitry, is licensed and marketed as Dolby Surround. In this domestic format the centre screen speaker is missing or is merely a sum of the left and right, which while reducing the hole-in-the-middle effect can also act to reduce the overall stereo effect. In Pro Logic Surround, the more effective active option, a discrete centre channel is offered, suitable for larger environments; it also offers reduced signal leakage between speakers and improved directionality. All systems offer a fixed time delay for the surround channel in the home environment; this will be between 25 and 30 milliseconds. Pro Logic 2 is a further development that produces five-channel surround – 5.1 (left, centre, right, left surround and right surround) – even from any stereo programme material, whether or not it is Dolby Surround encoded.

5.1 Surround Sound

Discrete five-channel audio is available domestically from many sources – cable television, satellite television, digital terrestrial television, high definition television, digital versatile discs, PC soundcards and so on. The standard encoding systems include MPEG audio and DTS, but the most widely known is Dolby Digital, which uses Dolby AC3 audio encoding technology. In an attempt to 'optimize sound for the particular listener', this decoder modifies the sound output according to use. Within the system data it:

● identifies a programme's original production format – mono, stereo, etc.;
● optimizes the dialogue level, adjusting volume;

- produces a compatible downmix from the multichannel sources;
- controls dynamic range, particularly from wide range cinema sound;
- routes non-directional bass to subwoofers as required.

Metadata

In broadcasting in the USA, the term metadata has a particular meaning, which has been standardized through the ATSC process. It applies both to 'packaged media' such as Dolby encoded material as well as transmitted programmes. As with Dolby Digital, the aim is to improve audio quality and facilities by offering the most suitable audio for the domestic environment – improving upon the standard broadcast NTSC audio transmission with its limited bandwidth and use of compression. (In Europe the higher bandwidth and short transmission paths create few audio problems.)

The areas metadata cover and the options provided are considerable. Among other facilities the metadata stream controls loudness (dialogue normalization) and compression. The gain is adjusted automatically by the receiver as the different programmes and their control data are received.

Viewers usually adjust their domestic volume levels to an acceptable level for speech reproduction. The system operates within these parameters. For example, the set is turned on and the transmitted voice-over level is set by the listener. The broadcast is actually 10 dB below peak, allowing 10 dB of headroom; the 'dialnorm figure' will be set to -10 dB. This programme is followed by a drama; here, the dialogue may be set at 25 dB below peak, allowing 25 dB of headroom for dramatic music. When the change of programme occurs, the dialnorm will turn up the level by 10 dB to maintain a constant level of speech across the change of programming.

Many other facilities are on offer in the ATSC specifications, including the option of various audio mixes and various different languages. The reduction of surround sound to two-channel stereo is also provided.

Video on the web

Producing video projects to be networked on the Internet is becoming a recognized method of distributing material – webcasting. Independent short films, celebrity interviews, film trailers, as well as business uses for training, recruitment and product launches, are all possible because of video streaming techniques. Some say 'What broadcasting entertainment was to the past, streaming content will be to the future', but much will depend on whether there is sufficient financial investment to improve picture quality.

Video streaming uses video compression techniques to produce moving images. The user downloads the appropriate streaming video (SV) player into his or her computer and clicks on the link, the material passes into a buffer stage and then the video stream appears a few seconds later. As the video progresses, more material is downloaded into the buffer. Although almost immediate, the quality is not perfect, often with freezed frames and fuzzy images; improving streaming technology and increasing bandwidths will improve the situation. However, good audio can compensate hugely for poor pictures.

Domestic video formats

Multichannel audio formats can be distributed by video cassettes or video disc and, in fact, over half the income from theatrically released films is derived from this type of domestic sales. Originally, audio quality on the domestic video recorder was very poor using analogue sound with narrow tracks at very slow speeds. However, audio frequency modulation recording improved the situation, producing good-quality hi-fi sound. Despite these vast improvements in sound quality, many consumers still listen on the linear tracks of non-hi-fi recorders. Those involved in audio post production should be aware of the inadequacies of video cassette sound and the vast differences in sound quality between hi-fi and 'normal' formats.

VHS format

The VHS (Video Home System) video cassette was introduced by JVC in 1976. The video recording system is similar to that used in the original professional U-matic system. The VHS format uses two audio tracks of 0.35 mm in width, each with a guard band of 0.3 mm. There is no guard band between the audio track and the tape edge; this makes the format susceptible to problems of poor tape splitting. Recording pre-emphasis and de-emphasis is employed at a tape speed of 1.313 inches per second (3.3 cm/second). Dolby 'B' noise reduction is available on many machines. On the standard analogue tracks, an audio bandwidth up to 10 kHz is possible.

VHS hi-fi

The poor quality of VHS analogue sound encouraged JVC to develop a high-quality sound system using AFM (audio frequency modulation) techniques. The VHS system modulates the tape using separate heads to those used in the video system, and a compression and expansion system. The audio is recorded below the video using a system called Depth Multiplex recording. Audio is first recorded on the tape from heads on the video drum. The video signals are then recorded over this, but only at the surface of the tape. The recorded FM signal is less susceptible to distortion or noise compared with the standard analogue tracks. The signal also tends to go into compression on overload, rather than violent clipping distortion – the perennial problem with digital recording. Analogue tracks are also provided to create compatibility with standard VHS tapes.

Mini DV

Also known as DV (originally DVC), this was the first digital video format available to the consumer in the mid 1990s. The maximum tape length is 120 minutes. Two tracks of high-quality, 16-bit audio are available, or four tracks of 12-bit audio. It is of near broadcast quality and used both domestically and professionally as a viewing format.

Optical video (laser) disc

Introduced in 1977, these were popular in Japan, where the video cassette did not gain acceptance. They were manufactured in a similar way to DVD-V discs, which are supplanting them.

DVD (digital versatile disc)

Introduced in 1997 in the USA and 1998 in Europe, this interactive audio/visual (a/v) disc produces high-quality multichannel digital sound and vision. Originally called the digital video disc, it very quickly became the digital versatile disc, a name indicating that it could become a universal carrier of a/v information for the home.

With its high capacity and better than CD audio quality, it is regarded as a catch-all format for music, films, computer games and data. The single-sided DVD can hold 2 hours of digital video, whereas a double-sided dual layer disc can hold 8 hours high-quality video or 30 hours of VHS-quality video. It can support widescreen and eight tracks of audio. In Europe, DVD uses MPEG-2 compression and PCM stereo sound; it supports Dolby Digital sound and discrete 5.1 multichannel audio. DVD-V (as it should be properly called) is part of a family of formats – DVD-R, DVD-RAM, DVD-RW as well as DVD-V.

DVD audio is recorded in 'layers'; access is via an interactive screen, to various parts of the programme. Musical DVDs include biographies of the musicians. Drama-based discs give information about authors, locations and even offer alternative storylines. These are produced in the same way as any other programme material, but each section is coded to be called up as required.

Within the home it is now possible to recreate the illusion of sitting in the cinema. More audio/visual material is available to view than there has ever been, but large-screen televisions and surround sound with interactive content may encourage fewer people to leave the comfort of their firesides and go to the cinema.

Glossary

AAF   Advanced Authoring Format. A new project exchange format still in development, which is intended to replace OMF.

Access time   The time taken to retrieve video or sound information from a store, which might be in the form of a video or audio tape recorder, a sound effects library, etc.

AC3   See Dolby AC3.

ADAT   Digital eight-track recorder using SVHS and tape.

A/D converter   A device that converts analogue signals to digital.

Address   A time or location within programme material, often selected as a 'go to' for machines via timecode.

ADL   Audio Decision List. Similar to an EDL, but used specifically to refer to an AES 31 project.

ADR   Automatic Dialogue Replacement. A system for re-recording dialogue in sync to picture, also known as looping.

AES   Audio Engineering Society.

AES 11   A standard for the synchronization of digital audio equipment in studio operations.

AES 31   An audio-only project exchange format.

AES/EBU   Digital audio standard for transferring 16/24-bit stereo audio between machines using a balanced XLR connector.

AFM   Audio Frequency Modulation. See Frequency modulation.

AIFF   (.AIF, .SND) Mac standard wave file format.

Ambience   (USA) See Atmos/Presence.

Amplitude   The maximum height of a waveform at any given point.

Analogue   (audio and video) A signal which is a direct physical representation of the waveform in, for example, voltage, magnetism or optical modulation. Digital audio replaces analogue by converting a signal to a series of numbers. Microphones and loudspeakers pick up and deliver 'analogue' sound waves.

Anechoic   Used to describe a space that has been treated to eliminate all reflected sound and that is acoustically 'dead'.

Aspect ratio   Television has an aspect ratio of 4:3 (1.33:1). This relates the width of the picture in relationship to its height; in high definition this is 16:9. In film, aspect ratio ranges from 1.65:1 to 2.55:1.

Artefact   This word is used to describe an unwanted digital distortion in picture or sound.

Assembly   The first stage of editing, when the various shots are joined together in a rough order to produce a rough cut.

Assembly editing   Video editing by adding one shot after another, each shot using its own control track. Therefore, a picture roll may take place at the edit. See Insert editing.

Atmos   Abbreviation for atmosphere sounds (UK), presence or ambience (USA).

ATRAC   Adaptive Transform Acoustic Coding. A digital audio compression system designed by Sony, used in photographic soundtracks and MiniDiscs.

Attack time   The time difference between the arrival of a signal into a compressor or expander, and the point at which the processor reacts to that signal.

Attenuate   To reduce the amplitude or intensity of a sound.

Auto-assembly/conforming   The assembling of an edited master from an edit decision list (EDL) using a computerized edit controller.

Auxiliary return   A line-level input into a mixing desk through which effects processors and outboard gear can be routed into the mix.

Auxiliary send   An output from the mixing desk, which can be used to all or part of a signal to effects processors or outboard gear.

AVR   Avid Video Resolution. A setting used by Avid that indicates the resolution at which video is digitized into the system.

Backing up   Saving a version of an edit to another location to prevent loss of data in the event of a system crash.

Back time   The calculation of a start point by first finding the finish point and subtracting the duration. This can relate to edit points or the recording of narration.

Balanced line   A method of sending a signal via two wires, neither of which are directly tied to the earth of the system. If an external electrical field is induced into the circuit leads, it is in-phase in both leads simultaneously. This unwanted signal will be eliminated by cancellation at the balanced input point, while the alternating audio signal passes through unaffected.

Bandpass filter   An audio filter found in mixing consoles designed to allow only certain frequencies to pass through it.

Bandwidth   Refers to the range of frequencies a device or system can record or reproduce (measured in hertz, Hz).

Baud   Unit for measuring the date of digital data transmission.

Bel   A relative measure of sound intensity or volume. It expresses the ratio of one sound to another and is calculated as a logarithm (a decibel is a tenth of a bel).

Betacam (SP)   A Sony trademark, an analogue professional ½-inch video format, often used to create worktapes used in TV and film post. SP stands for Superior Performance.

Bias   A high-frequency alternating current (up to 120 kHz) fed into an analogue magnetic recording circuit to assist the recording process.

Binary   Counting using only two digits as opposed to the usual 10.

Bit   Binary digit. Digital and PCM audio systems convert audio into on or off pulses known as bits.

Black burst   Also known as blacking, colour black and edit black. Provides synchronizing signals for a system to lock onto for stabilizing a videotape recorder.

Blanking level   The zero signal level in a video signal. Below this is sync information, above this visible picture.

Blimp   A soundproof cover to reduce the mechanical noise from a film camera.

BNC   British Naval Connector. A professional connector used mostly used in video and timecode applications.

Boom mic   Any type of mic held on a fixed or movable boom device such as a mic stand or fishpole.

Bouncing　Recording mixing, replaying and then recording again within the audio tracks of an audio or videotape recorder.

Breakout box　A unit attached to a system which has multiple output connectors that allow the system to interface with the outside world.

Burnt-in timecode　(Window dub, USA) Timecode transferred with picture from the timecode track of a videotape and visually burnt in to the picture as part of the image. This can only be used on copies for audio post production use. The burnt-in timecode is accurate in still frame, whereas the timecode figure from the longitudinal coded track is sometimes wrong.

Bus　A bus carries the signal output from a number of sources into specific inputs, e.g. mixdown busses sending signals to a multitrack recorder.

Buzz track　Ambience recorded to match the background of a scene. It can be a room tone or can contain voices – e.g. for a restaurant scene.

BWAV　Industry standard audio file format that also supports metadata.

Byte　By eight. A set of 8 bits, used in the measurement of the capacity of digital systems.

Camcorder　A combined video camera and recorder.

Cardioid　A type of mic term that has a heart-shaped frequency response pattern.

CAT 430　A background audio noise reduction device marketed by Dolby using their SR technology (replacing CAT 43).

Cathode ray tube (CRT)　The picture tube of a television monitor, visual display unit or phase meter, now being replaced by plasma screens.

CD　Compact Disc. Standard audio disc. CD-R denotes a recordable disc.

CD-ROM　Read-Only Memory. Contains only data rather than audio.

CD-RW　Read and Write. Not all CDs can play these.

Channel (strip)　An input on the mixing desk, which has its own fader, inputs, outputs and eq settings.

Chorus　A delay effect, which multiplies the original signal (e.g. a single voice into a chorus).

Clapperboard (clap sticks) A pair of hinged boards that are banged together at the beginning of a film double system sound shoot, to help synchronize sound and picture if timecode is not used. The clapper usually has an identification board as well.

Click track A sound with a regular beat used for timing in music scoring, which is produced electronically.

Clipping A term used to describe digital distortion created by excessive levels, which will result in the top of the waveform being chopped or cut off above a maximum level.

Clock The digital system that controls digital systems to ensure the digits stay together in synchronization.

Codec A coding/decoding algorithm that compresses data into a stream, e.g. MP3, Dolby E, etc.

Comb filter The effect produced by combining a signal with a slightly delayed version of itself.

Combined print A positive film on which picture and sound have been both printed.

Compression audio A method of reducing the dynamic range of sound using a compressor.

Compression data A method of reducing the data used in digital recording so that it takes up less space.

Compression ratio The amount of compression that has taken place, expressed in relation to the original signal – AC3 has a compression ratio of 12:1.

Console Colloquial term for an audio mixing device, called a 'desk' in the UK or 'board' in the USA.

Control track A regular pulse that is generated during videotape recording for synchronizing, it provides a reference for tracking control and tape speed and is found at the edge of the tape.

CPU Central Processing Unit. The main brain of a computer.

Crash What most systems do at some point; backup regularly to avoid loss of data.

Crossfade Describes a picture or audio transition where a fade out mixes directly into a fade in.

Crosstalk In multichannel equipment, crosstalk occurs when, for example, one channel of audio signal leaks into another. This term can also apply to electrical interference of any kind.

CRT print A video-to-film transfer using a CRT television screen.

Crystal lock   A system of driving a video or audio recorder at a known speed, at high accuracy, usually better than within one frame in 10 minutes. The timing synchronizing signal is derived from an oscillating quartz crystal.

Cut   Can be used to refer to a single edit (i.e. a straight edit without fades/crossfades). Also used to describe the various edited versions of a project (e.g. rough cut, fine cut).

Cutaway   A shot other than the main action added to a scene, often used to avoid a jump cut or to bridge time.

D-A, DAC   Digital-to-analogue converter.

Dailies   See Rushes.

Daisy chain   A series of devices (usually drives) connected via a single interface, where the audio signal must pass through each device in the chain before reaching the host system.

DAT   Digital Audio Tape.

DAW   Digital Audio Workstation.

dB   Decibel. A convenient logarithmic measure for voltage ratios and sound levels. One tenth of a bel, a 1 dB change is a very small but perceptible change in loudness. A 3 dB change is equal to double the power of the signal. A 6 dB change is the equivalent to doubling the perceived loudness.

Decay time   See Reverberation time.

De-esser   A device for reducing sibilant distortion.

Delay   Can be used to describe the delaying of an audio signal (measured in milliseconds), and can also refer to the actual hardware or software processor that can be used to apply delay effects such as flanging, echo and reverb.

Delivery requirements   A set of criteria for both finished picture and sound elements regarding mix format, tape formats, safety copies, etc. These must be delivered to the production company together with the masters.

Desk   (UK) See Console.

DigiBeta   (½-inch) Digital video format developed by Sony.

Digital I/O   An interface that transfers audio digitally between two systems.

Digitizing   Used to describe the recording of picture and sound into an NLE. Not used in relation to DAWs.

DIN   Deutsche Industrie-Norm. German industrial standard including the MIDI DIN plug specification.

Discontinuous shooting   Shooting video or film with one camera, moving the camera position for each change of shot.

Distortion   An unwanted change in a signal – the difference between the original sound and that reproduced by the recording machine. Distortion takes many forms, e.g. frequency, phase or digital.

DME   Dialogue, Music and Effects track, usually required as part of the delivery requirements.

Dolby   A family of audio system/noise reduction technologies.

Dolby AC3 and Dolby E   Data compression systems, which allow a multichannel mix to be compressed into a bitstream (AC1 and AC2 were earlier versions).

Dolby Digital (5.1)   A multichannel surround format comprising five full-bandwidth channels (Left, Right, Centre, Left Surround, Right Surround) and the '.1' channel, which is the LFE channel (carrying low-frequency effects only).

Dolby Pro Logic   This is the domestic version of Dolby SVA stereo sound; from a stereo encoded track (Lt Rt), four channels of audio, left, centre, right and surround, are produced. The centre channel offers several modes, normal, phantom and optional wide.

Dolby Pro Logic 2   A further dvelopment enabling a 5.1 signal to be electronically created from the stereo Lt Rt sources or even a simple stereo signal.

Dolby Stereo   This system encodes four channels of audio (Left, Centre, Right, Surround) into a stereo-compatible two-track Lt/Rt and fully decoded on playback.

Double system recording   Double system or separate sound recording. This is a production method that employs separate machinery to record sound and picture – it is used in film sound recording.

Drop frame   American system of timecode generation that adjusts the generated data every minute to compensate for the speed of the NTSC television system running at 29.97 frames per second.

Drop in   Making an electronic edit by switching from play to record while running.

Drop out   A momentary reduction or loss of signal, usually used with reference to tape or magnetic film.

DSP   Digital Signal Processing. Uses *processors* (such as expanders, limiters, compressors, etc.) and *effects* (such as delay, reverbs, pitch shift, etc.).

DTV   Digital Television. Can be broadcast through terrestrial, cable or satellite systems. In the USA it is defined by ATSC standards, with multichannel audio capability.

Dubbing mixer   (UK) Re-recording mixer (USA).

Dynamic range   The difference between the quietest and loudest parts of a signal. Also used to describe the range a component or system is capable of reproducing.

EBU   European Broadcasting Union. Produces standards for technical operations; particularly relevant is EBU N12, the technical standard for timecode, which is also known as EBU/SMPTE timecode.

Echo   The organized reflection of a sound from a surface.

Edited master   The final edited videotape with continuous programme material and timecode.

Edit points, edit in, edit out   The beginning and end points of an edit when a video programme is being assembled or soundtrack being recorded.

Effects   Sound effects, often abbreviated to fx.

Equalization (eq)   The boosting or decreasing of low, middle or high frequencies within an audio signal. Equalization is also used in tape machines to overcome losses in the recording processes. Equalization characteristics are recommended in the USA by NAB and in Europe by the IEC.

Error correction   Some systems have the ability to replace bits of digital data that are lost due to write errors on record.

Event number   A number assigned by an editor to each edit in an edit decision list.

Exciter lamp   The lamp in an optical sound reproducing machine, whose light is focused onto a photoelectric cell.

Expander   A type of processor that is used to increase the dynamic range of a signal below a user-defined threshold.

Fade in/out   Where a signal is ramped up or down to silence over a number of frames or sub-frames.

Fibre channel   A high-bandwidth serial protocol that is used in networking, which can transfer data at speeds of 100–400 MB/second.

Film speed   The universal film speed is 24 frames per second, used for theatrical productions. Other speeds are 25 and 30 fps.

Firewire (IEEE 1394)   A standard way of connecting apparatus to a computer for the high-speed transfer of data. Each manufacturer has their own different protocol, e.g. Sony iLink. Cable lengths have to be restricted to a few feet, but up to 63 devices can be controlled.

Flanging   A term used to describe two signals that are slightly out of phase with each other (sometimes used as an effect).

Foldback   A term used to describe the audio feed sent back to the performer(s) from the mixing console, which enables them to hear their own performance; particularly important for live music recording.

Foley   Creating sound effects by watching picture and simultaneously duplicating the actions.

Fps   Frames per second.

Frame   One film picture or one complete video scanning cycle, which is made up of two fields.

Frame accurate   Used to describe the ability of a system/software to maintain frame perfect sync lock.

Free run   The recording of timecode from a generator that runs continuously. A recording therefore has discontinuous timecode at each stop and start. See also Record run.

Freeze frame   The showing of one single frame of a videotape or film. See also Still in the gate.

Frequency   The number of times an event happens in a given time, e.g. frequency per second in sound (called hertz).

Frequency modulation   A method of encoding a signal whereby the programme material varies the frequency of a carrier signal – the level of the carrier remaining constant. Audio examples include stereo VHF radio; the hi-fi tracks of some domestic video cassette AFM recording give excellent quality.

Frequency response   The frequency range a component or system is capable of reproducing.

FTP   File Transfer Protocol. An Internet site that may be set up, to and from which audio files may be uploaded/downloaded.

Fx   Abbreviation for sound effects.

Gain   The extent to which an amplifier is able to increase/decrease the amplitude of a given signal – the ratio of input to the output level.

Gate   A processor used to increase dynamic range by gating or removing part of the signal when it drops below a user-defined threshold – often used to clean up dialogue that has a noisy background.

Gen lock   A system whereby all video equipment is locked to a common video synchronizing generator, allowing all to be synchronized together.

Gigabyte   The equivalent of 1024 megabytes (MB).

Glitch   A momentary break in sound due to data errors.

Group   Can be used to describe a fader which controls the output of a number of channels that have been 'grouped' together. Also used to describe a number of tracks that are grouped, so that any processing/editing applied to one track will apply to all.

Guide track   A soundtrack recorded that is likely to be replaced rather than used in the final mix, and that may be used as a guide in post production.

Haas effect   A psychoacoustic phenomenon where the mind identifies the first place from which a sound is heard as the origin of the sound, ignoring other sources of sound arriving a fraction of a second later (used in Dolby Stereo).

Handle   Additional frames added to the head and tail of each edit prior to transfer. These can be peeled out to improve a sound or picture edit at a later stage.

Hard drive (HD)   Data storage and retrieval device using fixed or removable magnetic disks that can be randomly accessed.

Hard effects   (USA) See Spot effects.

Hardware   The physical components that form part of an audio post production system – for example, the audio recorders, video player and computer.

HDTV   High Definition TV. A broadcast system at its best capable of producing film-quality pictures. In broadcasting has 720–1020 horizontal line resolution, supporting stereo and surround sound. The HD acquisition format is standardizing on 24p.

Head out   Tape is stored on the reel with the programme material head out – recommended for videotape storage.

Hexadecimal display   A facility for displaying 16 discrete characters 0–9 and A–F, as found in a timecode display.

High-pass filter   A filter that cuts out frequencies below a certain point, allowing frequencies that are higher to pass.

House sync   A central timing reference used to synchronize digital equipment and transports.

Hypercardioid   A more directional version of the cardioid mic.

Ingesting   A term that is beginning to be used to describe the importing of sound files into a digital editing system (as distinct from 'recording' or 'digitizing' in real time).

In point   The beginning of an edit, the first frame to be recorded.

Interlock   A term to indicate that two or more pieces of equipment are running in sync.

Intermodulation   The result of the interaction of two or more frequencies – the amplitude of one is modulated at the frequency of another.

I/O   Inputs/outputs, which may be analogue or digital and which allow a system to receive and send signals.

ISDN   Integrated Switched Digital Network. Used to send audio/sync signals over a phone line from one location to another. Often used for voice-over (commentary) recordings and ADR.

ISO (1)   International Standards Organization.

ISO (2)   Isolated Studio Output. The output of a single camera in a multi-camera set-up.

ITU 601   International standard for digitizing component video.

Jam sync   A process of locking a timecode generator to an existing coded recording so as to extend or replace the code.

Jog   A function on a video or audio device that allows image to be moved at any speed through stop to single speed with pictures forward and reverse (see Shuttle).

Jump cut   A cut from one shot to another, which breaks time continuity.

Kilo (k)   When used in digital systems it usually means 1024 (not 1000) – this represents the maximum value of a digital word of 10 bits.

LAN   Local Area Network. Used to transfer digital data from a client to a server – for example, from one workstation to another via Ethernet.

Latency   A delay measured in milliseconds that occurs when a signal is processed by a plug-in.

Laying sound   To place sound in its desired relationship to picture.

L cut (USA)   See Split edit.

Leader   There are different types: the most common are the Academy leader and the Universal leader. It gives a countdown to programme start, in film using flashing numbers, in video a clock. See Run-up/run-out.

Level   Volume of a sound. See also VU meter, Peak programme meter.

Level cut, straight cut   (Editorial cut, USA) Where sound and picture cut at the same point.

Level synchronization   A point where picture and sound tracks are in alignment (Editorial sync, USA).

Limiter   A limiter controls the peak levels of a signal above a user-defined threshold, usually to avoid distortion.

Lipsync   A term used to describe any on-camera speech and its sync relative to the picture.

Longitudinal timecode (LTC)   Timecode information recorded as an audio signal on a VTR or audio tape recorder.

Loop   Can refer to a single line of ADR, or can refer to a sound effect (usually an atmos) that is repeated or looped to run the length of a scene.

Lossy   Used to describe compression algorithms that remove data on encoding and do not restore it on decoding (e.g. MP3).

Low end   A term used to describe the bass elements of a signal (sometimes also called 'bottom end').

Low-pass filter   A filter that cuts out frequencies above a certain point, allowing frequencies that are lower to pass.

LTC   Longitudinal Timecode (sometimes called SMPTE code).

Lt Rt   Left total/Right total. The matrix-encoded two-track Dolby Surround mix that is decoded into its constituent channels on playback.

MADI (AES 10)   Multichannel Audio Digital Interface.

Married print   Sound and picture tracks married together on one piece of film.

Master   An original recording, video or audio, e.g. master edit, master music track, etc.

MB   Megabytes. A measure of computer storage equal to 1 048 576 bytes or 1024 kilobytes.

M&E   A Music and Effects version of a mix, from which all dialogue elements have been removed. Made for foreign sales.

Megabyte   The equivalent of a million bytes (1 048 576).

Metadata   A term used to describe the data that may accompany a sound file, such as slate number, source timecode, etc. Only supported by certain file formats.

MFX   Material Exchange Format. A file format that contains a completed project as streaming data. Similar in file structure to AAF.

Microsecond (μs)   One millionth of a second

Mic level   The output of a microphone that is amplified by a pre-amp before reaching the mic inputs of the mixing console or other device.

MIDI   Musical Instrument Digital Interface. A technology that allows an electronic device to interface with another, originally conceived for the music industry.

Millisecond (ms)   One thousandth of a second.

Mix   In sound, the combining of sounds. In vision, a dissolve.

Mixing console/desk   See Console.

Mono (monoaural)   A single track audio signal from a single sound source.

MOS, mute   Shot silent. From the German '*Mit Out Sprechan*' – without speech.

MP3   Compression format often used to deliver music over the Internet. Uses perceptual encoding to reduce data to a ratio of 1:12.

M/S   Mid(dle) and Side. A system for recording stereo sound using a cardioid forward directional microphone and a sideways figure-of-eight directional pattern microphone. Often used for sound effects recording, it produces good single-point pick-up and has good compatibility with matrix produced surround sound.

MTC   MIDI Timecode.

MTS   Multichannel Television Sound. An American system of stereo audio compatible with the MTSC system.

Multi-camera shooting   Using many cameras to cover a scene, all shooting at the same time; opposite of discontinuous shooting.

Multitrack recorder   An audio recorder with multiple recording tracks.

NAB   National Association of Broadcasters (an American organization).

Native File Format   A format which is the default file format of the system, rather than a format it can accept from another system in an adapted form.

NC   Noise Contour. A set of criteria used to assess the quietness of a room.

Near-field monitors   Speakers used in close proximity to the listener.

NLE   Non-Linear Editing system. Usually refers to picture editing systems.

Noise   Usually used to describe unwanted noise within a signal, either from a noisy location/set or from a system (system noise) through which the audio signal passes. Can be reduced by using noise reduction systems and noise gates during mixing.

Notch filter   A filter that is used to eq signals within a narrow range of frequencies.

NTSC   National Television Standards Committee. Analogue broadcast system primarily used in the USA and Japan, based on a 60 Hz mains system using 525 lines.

Octave   A musical interval spanning eight full notes. This is a 2:1 span of frequencies. Human hearing spans 10 octaves.

Off-line edit   The editing of material using non-broadcast-quality pictures, which are later conformed using on-line equipment.

Offset   The positioning of a device away from its normal synchronized position. The machines run continually with the offset in position. Offset is measured in time and frames.

OMFI   Open Media Framework Interchange. Developed by Avid to allow the open interchange of audio, video and edit information between different editing systems.

Omnidirectional   A mic whose frequency response pattern is equally sensitive through 360 degrees.

On-line edit   Refers to video or digital editing, where the material is being edited at full broadcast quality.

On-the-fly   Used to describe an action carried out whilst a system is still in playback or record, e.g. 'punching in on-the-fly'.

Oscillator   A device for producing pure-tone electric waves that are frequency calibrated.

Overmodulate   To modulate a recording beyond its recommended level and thus introduce distortion.

PAL   Phase Alternate Line: 625-line/50 Hz colour broadcast system used throughout most of Europe; often used to imply a 25 fps frame rate.

Parametric eq   An audio equalizer on which the parameters of bandwidth, frequency and the level of boost/cut in gain can be controlled.

Patching   To make a temporary connection between fixed cable terminals in a patch bay using a short length of cable called a patch cord.

PCM   Pulse Code Modulation.

PCMIA   Personal Computer Memory Card International Association. In effect, a laptop manufacturers; standard for memory storage, now used in some audio recorders.

Peak level   The highest (maximum) level in the waveform of a signal.

Peak programme meter (PPM)   An audio meter with a characteristic defined by the IEC, which registers the actual peak levels of a signal.

PEC direct switching (USA)   Also known as source playback monitoring, i.e. monitoring a pre-existing recorded mix and switching to compare it with the new mix to ensure levels are the same before re-recording. This is inserted using a punch-in record button – used in rock and roll systems. A term going back to photographic recording days.

Personal mic   A miniaturized mic that can be clipped or concealed within clothing.

Phase difference   The time delay between two time-varying waves expressed as an angle.

Phase scope   A form of oscilloscope having the ability to show phase relations between audio signals.

Phasing   Where the original signal is mixed with the same signal even slightly delayed, the resultant phasing will be heard as a clearly audible 'whooshing' effect.

Picture lock   The point at which no further adjustments to a picture are expected.

Ping-ponging   See Bouncing.

Pink noise   Random noise with a frequency spectrum that appears to be equal at all frequencies to the human ear. Its random nature makes it representative of actual programme conditions.

Pitch   A fundamental frequency that is perceived as 'pitch'.

Pitch shift/pitch changer   A hardware or software unit that can adjust the frequency of a signal without affecting its length.

Pixel   Acronym of picture element, the smallest units from which a picture can be built up.

Playout   An edit played out to tape, usually from an NLE, which is then used as a worktape (Avid refer to a playout as a 'digital cut').

Polarity   Positive or negative direction. The way of connecting a pair of signal conductors in phase or reverse phase.

Port   A digital I/O connection through which digital data can be imported/exported. A serial port sends or receives data bits in series (one after another). A parallel port can send or receive data bits simultaneously.

Post-fader   Used to describe a signal/device that is routed to or from the signal path after the channel fader.

Pre-fader   Used to describe a signal/device that enters the signal path before the channel fader.

Premix   A mix made of a number of tracks prior to the final mix – usually done to make the number of tracks more manageable.

Pre-roll   The time allowed before the start of a programme or edit to allow the equipment, video or audio, to come up to speed and synchronize (usually between 3 and 10 seconds).

Presence   See Ambience/Atmos.

Printing   A term used in film for copying or dubbing off picture or sound, e.g. printing a sound track, copying it.

Print through   A problem with archive material, through the transfer of one magnetic field from one layer of recording tape or film to the next on the reel. Pre-echo on tapes stored head out, post-echo on tapes stored tail out.

Production sound   See Sync sound.

Proximity effect   An increase in the bass element of a signal, which is emphasized as the distance between the mic and a sound source becomes less.

Pulse code modulation (PCM)   The standard way of recording digital audio.

Punch in/out   The selection of record on a video or audio system, usually on-the-fly.

Q   Used to describe the bandwidth affected by a filter – the lower the Q, the wider the range of frequencies affected by the filter.

Quantization   The process of turning a waveform into a number of set levels, essential for an analogue-to-digital conversion.

Radio mic   A type of personal mic that sends its output via an RF transmitter rather than a cable.

RAID   Redundant Array of Inexpensive Disks. A means by which drives are striped together to maximize I/O throughput, and offer system redundancy in the event of drive failure.

RAM   Random Access Memory. Temporary storage on a computer measured in megabytes or gigabytes.

R and D   Research and Development.

Real time   Any process that takes place in actual clock time.

Record run   The practice of running code only when the recorder, video or audio, is in record mode. The code is therefore sequential on the tape, despite the machine being switched on and off over a period of time. See Free run.

Red book   The data book that contains the specifications for manufacturers who make digital domestic equipment (CDs, DVDs).

Rendering   Used to describe the saving of processed audio as a new cue or event, usually in real time (also used to refer to video effects).

Release time   The amount of time a compressor takes to return to a passive state when the input signal is no longer above the user-defined threshold. The longer the release time, the longer it takes for a signal to be restored to its original level.

Reverb   A unit that artificially recreates natural reverbs for a variety of spaces (e.g. large hall, car interior). Also used to describe a setting on a reverb unit.

Reverberation   The persistence of sound in an enclosed space due to the summing of disorganized reflections from the enclosing surfaces.

Reverberation time   The time taken for a sound to drop to a level that is 60 dB lower than the original level (this measurement is referred to as RT60).

Room tone or buzz track   A recording of the background sound for a particular location/scene.

Rushes   Usually refers to the daily sound and picture material that has been developed in the lab overnight.

Safety copy   A copy of a master – to be used if the master is damaged.

Sample   A short audio cue recorded into a sampler.

Sample rate   The number of times a signal is sampled per second.

Sampling   The system used to measure the amplitude of a waveform so that it can be turned into a digital signal.

SAN   Storage Area Network. A high-speed data communications network that can be used to deliver audio, video and metadata to multiple devices or workstations.

Score   Original music recorded to picture for a programme or motion picture.

Scrubbing   Repeatedly moving sound or picture over a point or playhead to find a particular cue.

SCSI   Small Computer Systems Interface. A standard for the interface between a computer's central processor and peripherals such as additional hard disk drives. Up to 15 devices can be controlled simultaneously. Pronounced 'scuzzy'.

SD2 (or SDII)   Sound Designer file format native to Pro Tools.

SDTV   Standard Definition Television. This is a digital television format that offers 16:9 pictures (576 horizontal lines) with improved picture quality and MPEG digital sound.

SECAM   Sequential Colour with Memory. Television standard used in Russia and France, based on 625 lines/50 Hz.

Separate sound recording   See Double system recording.

Sequence   A sequence of edits assembled on the timeline of an NLE.

Sequencer   Software used to control a MIDI system.

Serial data   In timecode, code with all 80 or 90 bits in correct sequence, i.e. data in a continuous stream from one source.

Server   The sender in a Local Area Network.

Servo   An electronic circuit which, for example, controls the transport speed of a telecine machine. The servo may be adjusted by an external source.

Shoot to 'playback'   Playback of a soundtrack on location so that action may be synchronized to it.

Shuttle   Variable running of a video or tape recorder up to high speed from single speed.

Sibilance   A hissing consonant sound in speech most noticeable with the sounds of 's' and 'f'.

Signal path   Used to decribe the route taken by the audio signal through one or more devices.

Signal-to-noise ratio or S/N ratio   A term used to measure the background noise of a system.

Slate   An announcement, visual or aural, that gives identifying information for a recording.

SMPTE   The Society of Motion Picture and Television Engineers.

Snoop recording   A background recording left running throughout an entire ADR session (usually to DAT). Will be used in the event of drive failure, resulting in loss of the main recording.

Solo   A button on the channel strip of a mixing console, which allows one or more channels to be heard in isolation whilst all others are muted.

Source music   Music laid as though it is being played from an onscreen source within a scene.

S/PDIF   Sony/Philips Digital Interface Format (IEC-958). A 'consumer' version of AES/EBU, although it is found as an I/O option on many professional units.

SPL   Sound Pressure Level. The measure of the loudness of an acoustic sound expressed in dB: 0 dB SPL is near the threshold of hearing; 120+ dB is near the threshold of pain.

Split edit   An edit where the sound and vision start or finish at different times (L cut, USA).

Spot effects   Spot sounds that are not continuous and go with a specific action.

Spotting   Identifying picture points for sound effects or music cues.

Sprocket   The driving wheel or synchronizing wheel of a film apparatus provided with teeth to engage the perforations of the film.

Standard conversion   The converting of television pictures from one format to another, e.g. NTSC 525 lines/60 Hz to PAL 625 lines/50 Hz.

Stems   The constituent channels of a multichannel mix, e.g. the L/C/R/S/LS/RS/Sub stems for dialogue, music and sound effects, etc.

Stereo   A recording that takes place across two channels.

Sweetener   A sound effect designed to be used in conjunction with other sound effects to add texture and bass elements – not designed to be used by itself.

Thin   Colloquial term applied to sound lacking in lower frequencies.

THX   A set of technologies developed for the cinema, the studio and the home. In the cinema this relates to setting auditorium standards for the environment, the acoustics and the equipment – in accordance with SMPTE standard 202. In the studio THX PM3 standards set acoutic and monitoring standards, while in the home THX standards control loudspeaker and electronic design.

Time compression   A technique where the play speed of a recording is increased above the recorded speed. A pitch change device lowers the resultant pitch back to normal, resulting in a shorter actual time without any change in pitch.

Timeline   A graphical representation of the edit. This term is most used with reference to picture editing systems.

Tone   Continuous audio of one frequency often supplied at the beginning of a recording, which is used to line up input levels accurately. Usually of 1 kHz.

Tracklaying   The process of preparing or laying tracks in preparation for a mix.

Transient   A waveform with a sharp peak associated with sounds that have a hard 'attack'.

Transient response   The response of an amplifier or recording device to a very short, fast-rising signal. A measurement of the quality of a recording system.

Trim   To add or subtract time from or 'trim' an edit.

Triple track   Thirty-five-millimetre sprocketed magnetic film with three distinct soundtracks: dialogue, music and sound effects. Known as three stripe in the USA.

Twin track   Where the left and right channels of a signal actually contain two separate mono signals – often used for recording dialogue with two mics.

'U'-matic   An obselete ¾-inch video cassette format, occasionally used as a viewing format.

Undo   A command offered by some editing software, which will undo the last edit and restore the cut to its previous state – some systems have numerous levels of undo/redo.

Unity gain   Used to describe a situation where the input level and output level are the same.

Upcut (USA)   The loss of the beginning of an audio source, caused when an edit point is late, e.g. the loss of the first syllable of dialogue in a sentence.

USB   Universal Serial Bus. A common computer interconnecting system. Many devices use it, including recorders and MIDI systems. It can handle six streams of 16-bit audio.

User bits   Undefined bits in the 80-bit EBU/SMPTE timecode word. Available for uses other than time information.

Varispeed   A method of pitch changing achieved by speeding up and slowing down the replay machine.

Vertical interval timecode (VITC)   Timecode for video – encoded into the video signal, it can be replayed in tape format when the tape is stationary, since the video-head is rotating.

Virtual mixing   A technique that stores the mix purely as automation, rather than physically playing out the mix to another medium.

Voice-over (UK)   Explanatory speech, non-synchronous, superimposed over sound effects and music. Also called commentary or narration.

VU meter   A volume units meter that measures the average volume of sound signals in decibels and is intended to indicate the perceived loudness of a signal.

Wavelength   The distance between the crests of a waveform.

WAV file   Windows Wave File, a sound recording specific to a PC. A multimedia sound file that can be used within Windows with a variety of different soundcards and software.

White noise   Random noise with an even distribution of frequencies within the audio spectrum. This form of noise occurs naturally in transistors and resistors.

Wildtrack   A soundtrack that has been recorded separately from the picture, wild, without synchronizing.

Window dub   See Burnt-in timecode.

X-copy   A copy made as closely to the master version as possible.

XLR   A standard three-pin connector used for timecode or audio I/O.

Index